Tourism Studies and the Social Sciences

Tourism Studies and the Social Sciences is based upon a multi-disciplinary social science approach to understand the significance and role of tourism in contemporary society. It introduces social science disciplines to the reader and applies relevant theories to the understanding of tourism. Although each chapter addresses a particular social science discipline, the book includes extensive cross-referencing between the chapters to highlight the multidisciplinary nature of tourism research. A key theme of the book is how the economic and political structures of society influence the manifestation of tourism at a global level. It subsequently considers a variety of contemporary issues of society that transcend tourism, including citizenship and social exclusion; tourism as a form of trade; consumerism; the social, cultural and environmental consequences of tourism; feminism and ethics. The book is illustrated throughout with international examples and 'think points' to encourage the reader to reflect upon relevant theories and issues. Besides providing a wider understanding of tourism's role in society, it is also an aim of the book to make the reader think more widely about the society they live in. Each chapter includes:

- a brief introductory summary of the relevant discipline
- a critique of its main theories and concepts which have relevance to tourism
- a discussion of how the theories and concepts have been applied to tourism using cases and examples

Andrew Holden is Professor of Tourism at Buckinghamshire Chilterns University College, UK.

Tourism Studies and the Social Sciences

Andrew Holden

Routledge
Taylor & Francis Group

LONDON AND NEW YORK

First published 2005
by Routledge
2 Park Square, Milton Park, Abingdon, Oxon OX14 4RN

Simultaneously published in the USA and Canada
by Routledge
270 Madison Ave, New York, NY 10016

Routledge is an imprint of the Taylor & Francis Group

© 2006 Andrew Holden

Typeset in Times New Roman by
Florence Production Ltd, Stoodleigh, Devon
Printed and bound in Great Britain by
TJ International Ltd, Padstow, Cornwall

British Library Cataloguing in Publication Data
A catalogue record for this book is available from the British Library

Library of Congress Cataloging in Publication Data
Holden, Andrew.
    Tourism studies and the social sciences / Andrew Holden.
        p. cm.
    Includes bibliographical references and index.
    1. Tourism.   2. Tourism – Social aspects.   3. Social sciences.   I. Title.
    G155.A1H645 2005
    306.4′819–dc22   2005002007

ISBN10: 0–415–28775–8   ISBN13: 9–78–0–415–287753 (hbk)
ISBN10: 0–415–28776–6   ISBN13: 9–78–0–415–287760 (pbk)

CONTENTS

FIGURES

TABLES

BOXES

ACKNOWLEDGEMENTS

I will begin by thanking the Editorial Assistant at Routledge, Zoe Kruze, who has been a great help with the mechanics of this book. I would also like to give thanks to Her Majesty's Stationery Office for their permission to reprint Figures 1.6 and 1.7, to John Wiley & Sons for Figure 1.9, to Routledge for Figure 3.5, to Sage Publication Inc. for Figure 3.6, and to Blackwell Publishing for Figure 7.5. If any unintentional use of copyright material has been made in this book, the author would be grateful if the copyright owners could contact him via the publisher. Every effort has been made to trace the owners of copyright material.

My general thanks go to the colleagues and students I have had the pleasure of working with in the university system, some of whom have become good friends. In particular I would like to pay thanks to Raoul Bianchi, Peter Burns, Peter Mason, Julie Scott, John Sparrowhawk and Marcus Stephenson for discussing and developing perspectives on tourism. I want to pay thanks to my wife, Kiran Kalsi, for her support while writing this book. Finally, a special word for my dad, who, owing to illness, will unfortunately not be able to enjoy reading it.

INTRODUCTION

The emergence of tourism as an integral part of contemporary society has made it an area of interest to social scientists. The main aim of this book is to subsequently demonstrate how different social science disciplines can infuse the understanding of tourism. It provides vignettes of the social sciences, introducing the social science disciplines and theories that can be viewed as having relevance to tourism, in an attempt to make the links between tourism and the social sciences evident.

Besides being multidisciplinary, it is also interdisciplinary, as this approach is viewed by the author as being necessary to encompass a more comprehensive understanding of tourism. This is a view advocated by Graburn and Jafari (1991: 7) who comment: 'No single discipline alone can accommodate, treat, or understand tourism; it can be studied only if disciplinary boundaries are crossed and if multidisciplinary perspectives are sought and formed.' Subsequently, although the chapter headings of this book reflect single disciplines and areas of study in the social sciences, they should not be viewed as being rigid or impermeable. Different social science disciplines can lay claim to being the correct investigative approach to understanding tourism. However, the adoption of such a narrow and rigid approach negates the richness that different disciplines can give to a wider and more comprehensive understanding of tourism.

The understanding of tourism's role in society is the broad aim of this book. However, an equally important aim is to make the reader think about the society in which they live. Through providing an introduction to the social sciences and placing tourism within a wider context of issues facing society, the book goes someway to providing students with the analytical basis to interpret what is happening around them, and also provides a grounding in some of the major social issues that face them as citizens of the world at the start of the twenty-first century.

To date, perhaps with the exception of economics, the application of the social sciences to the investigation of tourism is relatively weak compared to other areas of social enquiry. This is partly because the growth of tourism as an activity on a global scale is a relatively recent phenomenon but also because tourism has frequently been viewed as an area of study that is frivolous and not appropriate for mature scholars. An objective of this book is to help to change this view and make future scholars think about the possible routes of social science research in tourism.

There are limitations to the book. In the view of the author it is unrealistic for any text about the social sciences and an area of enquiry, whether tourism or any other, to make claims to be definitive. The wealth and richness of material in the social sciences, and the growth in tourism literature, combined with the continuing revisions and insights in both areas, makes a definitive volume virtually unachievable. However, besides being informative the book will hopefully act as a catalyst for debate and raise students' interests in both the social sciences and tourism, with the aim of encouraging them to undertake further research and reading. An essential part of this process is the integration of 'think points' throughout the text. It will also hopefully encourage students to work as multi- and interdisciplinary scholars within the field of tourism studies.

The requirement for a multidisciplinary approach to tourism is further emphasised by the difficulty of trying to categorise areas of enquiry of tourism into singular social science disciplines. For example, if we talk of the tourism industry as being in the business of selling daydreams within a culture of consumerism, and of tourists fulfilling motivations and fantasies through participation in tourism, which of the social science disciplines of psychology, sociology or anthropology is the best or correct disciplinary approach to understanding this? In reality, all can enhance our understanding of tourism through their specific theories and methodological approaches.

Each social science discipline has its vested interests and its stake to be the authoritative social science of investigation of a particular social issue or phenomenon. Consequently, it is expected that this book will be open to criticism from different disciplinary perspectives. However, the author welcomes this in the context of promoting a stimulating and constructive debate about how the social sciences should be used to make sense of tourism.

The structure of the book is organised to reflect different routes of enquiry. The first part of the book proceeds to examine the relationship between the demand for tourism and society. Subsequently, a means of introducing the student to tourism, the first chapter has a geographical and historical basis, aiming to inform the student about the history and spatial growth of tourism. The general discussion in this chapter also highlights the multidisciplinary nature of tourism studies. Based upon the premise that most of the demand for tourism originates from developed urban societies, it continues to examine in Chapters 2 and 3 how the disciplines of sociology and psychology can be applied to help explain this pattern.

The book then proceeds to examine the economic and political factors that shape tourism within a global context. It aims to illustrate that where tourism develops it has not happened by chance, and that it is a function of economic and political structures as much as the workings of the market. Consequently, Chapter 4 analyses the economic aspects and theories of tourism, while Chapter 5 concentrates upon the political economy of tourism, which influences the goals of tourism and the distribution of its economic benefits.

The next two chapters of the book focus primarily upon the consequences and impacts of tourism. Chapter 6 introduces anthropology and how it can be applied to understanding tourism in both destination and tourism generating societies. Chapter 7 examines the effects of tourism upon the natural environment, using an area of the social sciences that has developed from a multidisciplinary basis itself; environmental studies. Chapter 8 focuses on emerging areas of the social sciences that are in their very early stages of application to tourism. Rather ironically, the first part of this chapter turns to an area of philosophy that predates the social sciences, that of ethics. The second half of the chapter examines the application of feminist studies to tourism.

The social sciences

Having emphasised the need for a multidisciplinary approach from the social sciences to enhance the understanding of tourism, it is also necessary to discuss the meaning and scope of social sciences, which is not agreed upon and remains a matter of philosophical debate. A broad definition of the social sciences is that they may be viewed as disciplines that aid us in understanding the processes and patterns of human behaviour and the societies that they create (Hughes, 1990). McLeish (1993: 688) comments: 'The social sciences are a study of disciplines concerned with the study of human behaviour.' McLeish (ibid.) includes as the 'standard' social sciences: sociology, anthropology, economics, political science and psychology, but adds that the sphere of influence includes other disciplines such as history, philosophy and geography and more recent areas of study such as feminism and ecology. However, the contentious nature of what constitutes the social sciences is exemplified by the discipline of history as McLeish (1993: 351) observes: 'History may be the queen of humanities but it is the bastard child of the social sciences.'

Trigg (1985) suggests that the disciplines of sociology, social anthropology, politics and economics would inevitably be viewed as social sciences. To this list he also adds history, on the basis that it studies the interaction of humans in society, even if in the past. The *Annals of Tourism Research*, in its special issue of 'Tourism Social Science' published in 1991, included anthropology, ecology, economics, geography, history, politics, psychology and sociology under the aegis of the social sciences.

Hughes (1990) suggests that what the social science disciplines have in common is that they have all been influenced by the natural sciences in their quest to understand human behaviour, hence the desire to create sciences of human behaviour. However, it is a matter of debate whether human behaviour can be studied in the same way as the phenomenon of the natural world. For example, while sociologists can be said to study the social world, the philosophical battle of how to understand the social world remains a lively one. Broadly, this philosophical battle depends upon whether society and human behaviour can be measured and theories constructed in a similar way to the natural sciences such as physics and chemistry.

Yet, the social sciences lay claim to an empirical base, which Trigg (1985) argues is to an extent a legacy of the success of modern physical sciences in shaping and influencing the world and the path of human development. As Okasha (2002) points out, the advancement of the natural sciences, e.g. physics, chemistry, biology, vis-à-vis the social sciences, has led to calls for social sciences to 'ape' the methods of the natural sciences. However, Trigg (1985) also observes that the pressure to conform to the one model of empiricism is being increasingly scrutinised as cynicism has increased in the latter half of the twentieth century about the ability of science to further human welfare.

Therefore, a key question facing social scientists is whether laws exist in the social world to govern human behaviour in the same way as laws, e.g. gravity, exist in the natural world. Is there, for example, a set of laws that can be discovered to explain tourism as a phenomenon? Is the aim of the social sciences to unearth the laws that govern human behaviour and explain its causes? Certainly, as will be explained in Chapter 2, the early traditions of sociology fell very much into this category, seeking to establish rules through a naturalist philosophy. As Trigg (1985) emphasises, it is *naturalism* that has historically had most influence upon the social sciences. Within 'naturalism' as Trigg (1985: 3) points out: 'The scientific character of the social sciences is emphasised, and anything that cannot be subsumed under scientific laws is excluded.' Hence this philosophy is based upon an empirical and positivist approach of establishing hypotheses and the measuring of social characteristics. Highly influential in advocating the philosophy of naturalism to understand the social world in the first half of the twentieth century was a group of philosophers and mathematicians known collectively as the 'Vienna Circle', which was centred upon Vienna University (Crystal, 1994).

An alternative approach to naturalism is *humanism*, which rejects the notion that empiricism is capable of answering all the questions about our place in nature and the world. Humanism stresses that people are different from physical objects and consequently must be understood differently. Subsequently, the naturalist approach of taking the natural sciences as a model to follow for understanding the social world is rejected.

It is consequently a matter of debate whether 'social' qualifies science in the same way that 'natural' or 'physical' do (Trigg, ibid.). Not only is it contestable whether human behaviour can be studied in all the same ways as the phenomena of the natural world, but, also, it is contentious whether 'science' is in any sense the correct term for all these disciplines. Adherents once claimed it was, arguing that any form of study using a scientific method of hypothesis, experimentation and analysis was a science. This was certainly the view in the eighteenth and nineteenth centuries in Europe, when many of the disciplines were established. However, as the disciplines have burgeoned and begun to interpenetrate one another, sharing research techniques and combining evidence, there has been a strong impetus towards counting the whole study of human behaviour as a single entity and calling it 'social studies'.

In this philosophical debate that underpins the social sciences, the key issue is about the nature of evidence, or as Hughes (1990: 4) puts it: 'how we know certain things, believe others, how we know things to be true or false, what inferences can legitimately be made from various kinds of experiences, what inferences consist in, and so on'. The vast difference in the philosophical approach between naturalism and humanism raises key questions about human assumptions of making sense of our social world. The questioning of assumptions is known as an *ontological* issue. Probably the most basic ontological question in the social sciences is whether 'the reality' to be investigated is external to the individual, that is imposing itself on individual consciousness. As Phillimore and Goodson (2004: 34) put it: 'Ontology is the study of being, and raises questions about the nature of reality while referring to the claims or assumptions that a particular approach to social enquiry makes about the nature of social reality.' In other words, do social laws and reality that govern our behaviour exist independently of the human consciousness or are social laws the product of one's mind?

Linked to this ontological issue is the debate over the 'nature of knowledge', termed *epistemology*. For Hughes (1990), epistemology is concerned with the character of our knowledge of the world and the justification of the claims about the way in which the world can be made known to us. As Burrell and Morgan (1979) point out, *epistemological* concerns focus upon the grounds of knowledge, about how one might begin to understand the world and communicate this knowledge to fellow human beings. This includes assumptions about what forms of knowledge can be obtained and how one can sort out what is to be regarded as 'true' from what is being regarded as 'false'. With relevance to the epistemology of tourism, Tribe (2004: 46) states: 'Knowing about how and what we know in tourism is an epistemological question, epistemology being that branch of philosophy which studies knowledge.' A key question is, is it possible to identify and communicate the nature of knowledge as being hard, real and tangible? Alternatively, is knowledge a softer, more subjective and spiritual experience based upon the insight of an essentially personal nature?

Associated with ontological and epistemological issues, but conceptually separate from them, is a third set of assumptions concerning *human nature* and in particular the relationship between human beings and their environment. That is: (i) are human beings and experiences products of their environment, to which human beings respond in a mechanistic or even a deterministic fashion to the situations encountered in their external world; or (ii) should 'Free will' be given the centre stage, in which we create our own environment, 'the master rather than the marionette'.

In these two polarised views of the relationship between human beings and their environment we are identifying a great philosophical debate between the advocates of *determinism* on one hand and *voluntarism* on the other. If one treats the world as a hard external, objective reality then the focus is upon an analysis of the relationships and regularities between the various elements it compromises. The methodological issues of importance are thus the concepts themselves, their measurement and the identification of underlying themes. This perspective expresses itself most forcefully in a search for universal laws that explain and govern the reality that is being observed, such as tourism.

In the alternative view of voluntarism, stressing the subjective experience of individuals in the creation of the social world, the principal concern is an understanding of the way in which the individual creates, modifies and interprets the world in which he or she finds themself. However, if we can accept the validity of the different philosophical approaches to our understanding of the world, and the use of a variety of research methods that reflect these philosophies, it is possible to develop theories about our observations and patterns of events that surround us.

According to Babbie (1995), theory has three functions for research. First, if we know why something happens we can anticipate whether and what will make it happen in the future. Second, through our understanding of why events occur, we can control and influence these events. Last, theories can direct research efforts, indicating the likely direction of empirical discoveries. Thus, through the use of theories the frontiers of our understanding of a phenomenon such as tourism can be developed further. However, as will become apparent while reading this book, there is a limited theoretical development in tourism studies. Not least, this is a consequence of the relatively limited attention compared to other phenomena of society it has to date received as an area of study from social science disciplines. Tourism as an area of study can therefore be said to be 'youthful' and 'relatively immature' (Tribe, 1997). As Tribe (2004) suggests, tourism

?agree?.

is perhaps best thought of as an area of study rather than as a discipline, because although tourism utilises a number of theories such as motivation, economic multipliers and development theories, these are the property of other social science disciplines and areas of study.

Jafari (2001) views the development of theory in tourism as having passed through four evolutionary stages or platforms of knowledge. Economists dominated the 'advocacy' platform in the 1960s that emphasised a strong economic rationale for tourism. The 'cautionary' platform emphasised the significant economic, cultural and environmental disadvantages of tourism, as the periphery of mass international tourism was extended to less-developed countries during the 1970s. The 'adaptancy' platform emphasised alternative forms of tourism to the model of mass tourism, which were more environmentally and culturally responsive to the needs of destination communities, a theme which became popular during the 1980s. The 'knowledge' platform is a reflection of the growing maturity and importance of tourism as an integral part of society, and that the focus on the impacts and development of tourism in the first three platforms offers a limited view of tourism. Consequently there is a requirement to develop a comprehensive knowledge of tourism based upon objectivity and scientific investigation. This book subsequently explores the linkages between social science theories and tourism in an attempt to move towards a more comprehensive knowledge.

A HISTORICAL GEOGRAPHY OF TOURISM

1

This chapter will:

- describe patterns of contemporary tourism;
- give a historical account of tourism;
- consider how tourism can be thought of as a system;
- analyse the social and economic changes in place that have transformed society from a largely static one to a mobile one;
- emphasise the role of the Industrial Revolution in the development of mass participation in tourism.

Introduction

Figure 1.1, taken in Cannes in the south of France, represents the type of view that most people would probably associate with tourism. Yet, if we were to analyse the photograph more deeply, we could begin to ask a range of questions. For example, these could include: why is a central focus of the photograph based upon one person?; where do the people on the beach come from?; why do they choose to spend their time in this type of surroundings?; what are their personal motivations for behaving this way?; how do they interact with each other?; how does their presence impact upon the local economy?; what are the consequences of their behaviour upon the nature and local culture of the area?; and how does this experience influence their perceptions of life in their home societies?

To attempt to begin to find the answers to these questions it is necessary to treat tourism as a subject for academic inquiry. As the term 'tourism' has become increasingly omnipresent in the global vernacular, it has brought with it economic, social, cultural and environmental changes, emphasising the requirement for a deeper understanding of this phenomenon of contemporary society. However, while increased participation in tourism has been encouraged by rising material standards of living, media exposure, a mature tourism industry and the desire of people to visit different places away from their home environment, the majority of the world's people remain excluded from participating in tourism, certainly in its popular western connotation as holiday or vacation time free from the regular routines of work and home.

Figure 1.1 Focusing on tourism; Cannes in the south of France.

Nevertheless, the increase in the rate of participation in tourism since the 1950s has been dramatic as is shown in Figure 1.2. In 1950 there were approximately 25 million international arrivals recorded in the world compared to over 700 million in 2003, and by 2020 it is forecast that this number will have risen to over 1,600 million (WTO, 2003a). However, tourism is not just about the movement of people between countries, it also involves the movement of people between places within countries, referred to as 'domestic tourism'. Although a global estimate of the amount of domestic tourism is difficult to attain, its volume is likely to greatly exceed that of international tourism.

While the number of people participating in tourism may be rapidly growing, the spatial distribution of tourism around the world is not uniform. Figure 1.3 shows the distribution of international tourist arrivals by geographic region.

The dominance of Europe as a destination for tourism is evident, accounting for over half of all international arrivals. A range of interacting factors explains this pattern, which will be expanded upon in this chapter. These include the effects of the Industrial Revolution and economic development; historically high standards of living compared to other countries of the world with the exception of North America and Australasia; the political geography of Europe comprising of geographically small nation states; the freedom and ease of movement as a consequence of the creation of the European Union; a highly developed air, road and rail structure that facilitates international travel; and the comparatively early development of a tourism industry.

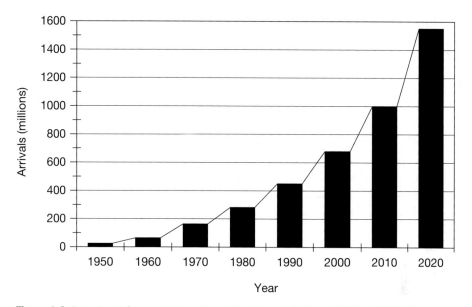

Figure 1.2 Actual and forecast international tourist arrivals from 1950 to 2020 (data source: WTO, 2003b; WTO, 2004a). *Check stats for 2005/6.*

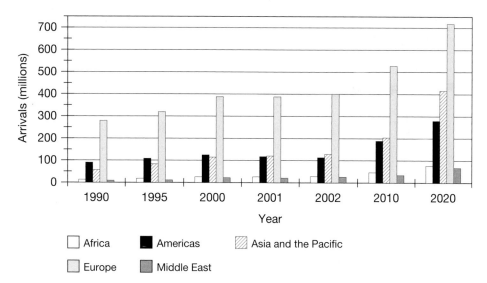

Figure 1.3 International tourist arrivals by region (actual and forecast) (data source: WTO, 2003).

As the WTO (2003c) point out, tourism flows are unevenly distributed between various regions of the world, with Europe, North and South East Asia, and North America accounting for an overwhelming proportion of both supply and demand. Around 58 per cent of all international arrivals take place in Europe, 16 per cent in North and South East Asia and around 12 per cent in America, representing almost 9 out of 10 international arrivals. Most of this international travel takes place intra-regionally, with approximately 87 per cent of all international arrivals in Europe originating from Europe itself; the corresponding figure for the Asia-Pacific region being 77 per cent; and the Americas being 71 per cent (WTO, ibid.).

Expressed in terms of the flows of tourists, there are six major patterns that account for nearly a quarter of all international tourist arrivals. Ranked in terms of the approximate numbers of arrivals, these are: (1) Northern Europe to the Mediterranean (120 million); (2) North America to Europe (23 million); (3) Europe to North America (15 million); (4) North East Asia to South East Asia (10 million); (4) North East Asia to North America (8 million); and (5) North America to the Caribbean (8 million) (WTO, ibid.).

A critical factor to explain the flows of tourism between Northern Europe and the Mediterranean, and between North America and the Caribbean, is the attractiveness of the climate of these destination regions. As Williams (1997) points out, it is the summer beach and the sunshine holiday that have propelled the Mediterranean into pole position in world tourism. The flow from North East Asia to South East Asia also has a large sun, sea and sand component, although business travel and visiting friends and relatives (VFR) are additional key components of this market (WTO, 2003c). Similarly, the flows of travel between Europe and North America, and between North East Asia and North America, encompass a variety of types of traveller.

In terms of the demand for tourism, the importance of economic development for generating tourism is emphasised by consideration of the expenditure upon tourism by individual countries, as is shown in Table 1.1. Ranked in order based upon total expenditure for 2002 and 1990, two trends are evident. The first being the dominance of economically advanced countries in the top six places; the second being the growing prominence of those countries with rapidly expanding economies, notably China and to a lesser extent the Russian Federation. In 1990 China was ranked only fortieth in terms of tourism expenditure; by 2002 it had risen to seventh place. Influential factors in this increased expenditure on tourism are the rapid economic growth in China during the last decade and also moves to reduce the political restrictions on foreign travel. Similarly, the Russian Federation has moved from twenty-third position in 1990 to tenth in 2002. The opening of the Russian economy to market forces and the lifting of restrictions on travel after the collapse of communism in 1991 are major factors to explain this growth in expenditure on international tourism.

Through considering spatial patterns of contemporary tourism it is perhaps becoming evident that where people go to and why they choose to go there is not determined by chance. Already the influence of levels of economic development and climate on patterns of tourism has been noted. However, these are far from being the only factors that shape patterns of contemporary tourism.

The emphasis of this chapter is subsequently to explain how contemporary patterns of tourism have emerged over time. To aid this exploration, a synergy is made between the disciplines of geography and history, as there is a traditional link between the two, with historians having studied particular time periods and geographers particular places (Johnston, 2003). Both offer exploratory accounts of places and periods through a synthesis of available material, and therefore a historical geography of tourism will enable a better understanding of the factors that have and continue to influence patterns of tourism.

Table 1.1 *World's top spenders on international tourism in 2002 and 1990 (sources: WTO, 1998; WTO, 2004b).*

Country	Actual expenditure on international tourism in 2002 (US$ billion)	Rank in 2002	Rank in 1990
United States	58.0	1	1
Germany	53.2	2	2
United Kingdom	40.4	3	4
Japan	26.7	4	3
France	19.5	5	6
Italy	16.9	6	5
China	15.4	7	40
Netherlands	12.9	8	9
Hong Kong (China)	12.4	9	N/A
Russian Federation	12.0	10	23

N/A=Not available

Think point

Which countries have you travelled to? How do your experiences of travel and tourism compare to your parents and grandparents experiences of travel and tourism when they were your age?

Thinking of tourism as a system

When thinking of tourism at its most simplistic level it involves a spatial separation between 'home' and 'away' and travel between these two zones. Yet this seemingly simple act carries with it a range of requirements and consequences, in terms of the services required to meet the needs of tourists and the impacts tourists have upon places. These characteristics are illustrated by the following definition of tourism from Jafari (1977: 8) as: 'a study of man away from his usual habitat, of the industry which responds to his needs, and the impact that both he and the industry have on the socio-cultural, economic, and physical environments'. Similarly, Mathieson and Wall (1982: 1) comment:

> The study of tourism is the study of people away from their usual habitat, of the establishments which respond to the requirements of travellers, and of the impacts that they have on the economic, physical and social well-being of their hosts.

Emphasising the importance of the concept of place in tourism, Pearce (1995: 20) suggests:

> Tourism is essentially about people and places, the places that one group of people leave, visit and pass through, the other groups who make their trip possible and those that they encounter along the way. In a more technical sense,

What about ports , how do these groups

tourism may be thought of as the relationships and phenomena arising out of the journeys and temporary stays of people travelling primarily for leave or recreational purposes.

It can also be added to this definition that besides travel for recreational purposes, as has already been referred to in the context of major tourist flows, other types of tourism include: business; visiting friends and relatives; and religious pilgrimage.

The theme of the connections between places is also emphasised in Tribe's (1997: 64) definition of tourism as: 'The sum of the phenomena and relationships arising from the interaction in generating and host regions, of tourists, business suppliers, governments, communities, and environments.' Both these latter definitions of Pearce and Tribe, unlike the previous two, emphasise the generating regions or the home societies of tourists as an integral part of the analysis of tourism. Tribe (1997) also suggests a number of other areas that are important in moving towards an understanding of tourism, including issues of tourist motivation and choice; host community perceptions of tourism; economic, cultural and ecological impacts; the measurement of tourism; and policy and planning.

This latter definition also indicates the wide variety of groups that have an interest in tourism, including the tourist; local communities; governments; and the tourism industry, all of whom may be referred to as the 'stakeholders' in tourism in the sense of seeking various benefits from being involved with it. Subsequently, it is apparent that tourism would seem to involve more than just the solitary act of travelling. By travelling a variety of requirements are generated, including at a most basic level a requirement for transport and accommodation, while a range of economic, social, cultural and environmental impacts are also generated.

Instead of thinking about tourism as something that exists externally to the individual, an alternative way to think about it is how the individual constructs and gives meaning to it. Such an experiential definition of tourism is given by Franklin (2003: 33) who defines tourism as: 'an attitude to the world or a way of seeing the world, not necessarily what we find only at the end of a long and arduous journey'. In this definition emphasis is subsequently placed upon individuals constructing their own meaning of tourism instead of it existing as a defined or real entity.

One approach to understanding tourism that attempts to encompass these different elements is to think of it as a system. Towner (1996) suggests that by thinking of tourism as a system we can interpret it as being dynamic, open to change and continuity, and that it can be viewed to operate on any spatial scale. This is the approach that is favoured in this book, not least because it allows for a holistic interpretation of tourism, utilising various social science perspectives. Page (1995) adds that the advantage of a systems approach is that it allows the complexity of the real-life situation to be accounted for in a simple model, demonstrating the inter-linkages of all the different elements. One way of visualising the inter-linkages of tourism is to think of it as a spider's web, with each part inter-connected to each other, so that by touching one part ripple effects are felt throughout the system (Mill and Morrison, 1992). Importantly, this necessitates a need for responsiveness in a specific part of the system to changes in another part (Mill and Morrison, 2002). A suggested model of the tourism system and its relationship with the social sciences is shown in Figure 1.4.

perceive the place they will be visiting? Does it match their expectations + how does FEd Manage' meet their expectations.

A HISTORICAL GEOGRAPHY OF TOURISM   13

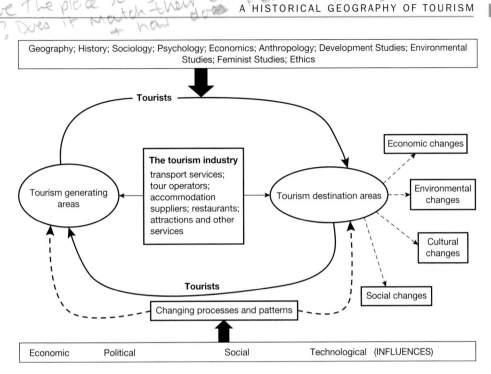

Figure 1.4 The tourism system and the social sciences.

This model attempts to integrate both a historical and dynamic perspective into the explanation of the tourism system by emphasising the changing political, economic, social and technological processes and patterns in societies that influence tourism. It emphasises that although the demand for tourism may be viewed as a consequence of changes in generating societies, it also brings changes to the societies of destinations that tourists visit. These economic, environmental, cultural and social changes are shown on the right-hand side of the model and are commonly collectively referred to as 'impact' issues in the tourism literature. These impacts may also be experienced in the societies where tourists come from, for example tourists have various types of experiences when visiting destination areas, the memories of which they carry back with them and reflect upon after their return home. Consequently, tourism has the potential to influence people's attitudes towards their own societies.

Similarly, the economic impact of expenditure on tourism is not restricted solely to destinations, as the use of tourism industry suppliers such as airlines and tour operators also generates extra economic demand and employment in the home societies of tourists, while tourist expenditure economically influences a country's balance of payments as is discussed in Chapter 4. Environmental impacts are also caused in generating regions, for example through the construction of airports, and associated increased levels of vehicle and air traffic. Increasingly societies in the world have the characteristics of what Pearce (1995) refers to as 'reciprocity', having functions as both tourism generating and receiving areas. Tourism can subsequently be viewed as a complex and dynamic system, involving reciprocity by bringing changes to places where tourists both go to and come from.

The degree of change attributable to tourism in a particular place will depend upon its economic and social characteristics. Tourism destinations vary considerably in their characteristics, for example cosmopolitan cities or indigenous villages with little contact with other cultures may become tourism destinations. It is also important to remember that tourism is just one factor of change in an era of 'globalisation' in which closer economic dependencies between nations, electronic media and information technology are also major determinants of social change.

The vital link in terms of facilitating travel between places is the tourism industry. Without the travel services provided by airline, railway, passenger ship and coach companies, the patterns of contemporary tourism would be very different. Suppliers of accommodation ranging in scale of provision from large multinational hotel chains such as Sheraton to individual family-run lodgings are also essential for tourism. We need to think, too, of some of the other services required by the typical tourist such as restaurants and other types of food and beverage outlets; souvenir shops and retail outlets; and amusement and theme parks; to begin to comprehend the diverse range of services provided by what may be loosely termed the 'tourism industry'.

Integral to the model shown in Figure 1.4 is an emphasis upon the human dimension of the tourism system. In both tourism generating and destination areas, people are a key focus of the system. To understand the tourism system we consequently need to utilise various social science disciplines that can give us a greater understanding of people, their behaviour and impact. Subsequently, the disciplines of psychology and sociology have a central role in helping us understand why people choose to become tourists. For tourism to take place also requires the decision to allocate resources to it, by individual consumers, governments, entrepreneurs and business. The understanding of the mechanisms of resource allocation can be aided by the discipline of economics, while a comprehension of the wider political and economic decisions to develop tourism can be analysed through development studies. In attempting to understand the impacts of tourism in destination areas, the use of anthropological theories has an important role to play.

To aid our understanding of the environmental impacts of tourism, it is necessary to engage with a more recent area of study to have emerged within the social sciences in response to an evolving and changing society, that of environmental studies. Similarly, in terms of a more recent area of social theory, feminist studies can aid our understanding of how tourism is influenced by and influences the power relationships between men and women. Finally, given the many issues that revolve around tourism, any discussion of tourism would be incomplete without reference to ethics. Although strictly a branch of philosophy rather than the social sciences, the ethics of tourism stakeholders is highly topical and relevant to the future of tourism and hence needs to be addressed by students of tourism.

Think point

How would you explain to someone what tourism is? What kind of impacts have you observed as a consequence of tourism? Have any experiences you have had as a tourist changed your view of your home society upon your return?

An introduction to geography and history

From the preceding description of patterns of contemporary tourism, it is evident that tourism is based upon the movement of people between places and the consequent use of space. The social science discipline that links the concepts of place and space is that of geography (McLeish, 1993). The origins of the term 'geography' can be traced to approximately 300 BC when it was used by Greek scholars in Alexandria (Holt-Jensen, 1999), with *ge* meaning 'Earth' and *graphia* or *graphos* meaning 'drawing', subsequently placing 'an emphasis upon a description of the Earth' (McLeish, 1993: 314). The ancient roots of geography are also apparent through the work of Ptolemy (AD 90–168) in his book of eight volumes known as Ptolemy's *Geography*. The primary focus of these works is upon calculations to measure the dimensions of the earth; estimates of the latitude and longitude of 4,000 places; and maps of different parts of the world (Holt-Jensen, 1999).

Subsequently, 'topographic' descriptions through map-making, surveying, mathematical and geometrical enquiry, and exploring, can be viewed as representing the roots of geography (Collins, 2004). Although today the term topography is usually associated with the physical characteristics of land, in ancient times topographical description meant a more general description of places including cultures (Holt-Jensen, 1999), a role that is now undertaken primarily by anthropologists. A notable early contributor to the topographical description of the world's natural conditions and cultures was the ancient Greek Herodotus (*c*.485–425 BC), who was also a noted early historian, and whose writings included references to tourism. Herodotus spent much of his life travelling and recording the observations of his travels, visiting Egypt, Syria, Persia, Asia Minor, Sicily and Babylon, subsequently being widely recognised as the first travel writer (Sharpley, 1999).

In terms of geography as a subject for academic study, Johnston (2003) comments that elements of geography were being taught in the ancient British universities by the late sixteenth century. Reflecting its ancient origins, it combined the technical mathematical skills of navigation and map-making with literary and descriptive skills based upon numerous written accounts of the flora, fauna, landscapes and people of distant origins. However, geography only became recognised as a distinct discipline within separate university departments in the UK and the US in the early twentieth century.

Today two broad categorisations of geography can be recognised: physical and human geography. During the twentieth century geography also became increasingly influenced by other social sciences, incorporating historical, economic, social, political and cultural geographers. Although geography may seem diverse in its range of enquiry, its purpose is explained by Holt-Jensen (1999: 5) as follows:

> Geography exists to study variations in phenomena from place to place, and its value as an academic discipline depends on the extent to which it can clarify the spatial relations and processes that might explain the features of an area or place.

Similarly, Castree (2003: 166) comments: 'the discipline of geography is still very much about the study of the world's variable character – and thus still very much alive and well.' However, in Castree's view, regional geography, i.e. treating areas of the world as being separate entities, can no longer account for these variations. Today, places of the world are increasingly connected economically, through information technology and media, and by travel. Consequently, economic and social changes in one place are likely to affect other places. Similarly, the approach taken in this book is that we cannot understand tourism out of the context of changes in places and their inter-connectivity though space. As Lew (1999: 1) comments: 'And place is, also an intrinsic element of tourism, as all tourism involves some form of relationship between people and places that they call "home" and "not home".'

Before proceeding to a historical and geographical account of tourism it is necessary to first consider the role of history in an enquiry of tourism. Similar to geography, the origins of history can be traced to the Ancient Greeks, notably Herodotus (c.485– 425 BC), already referred to as an eminent figure of ancient geography. The word *historia* means 'knowledge discovered by inquiry' (McLeish, 1993), however, it was not until the end of the nineteenth century that history became an academic discipline within universities. Before this period, history had largely been regarded as both storytelling and as a source of moral guidance, stories often reflecting the achievements and follies of monarchs, political elites and religious saints (McLeish, ibid.).

Although the dominant philosophy of history became during the nineteenth century the one advocated by the German historian Ranke to emphasise 'facts', or 'simply to show how it really was (*wie es eigentlich gewsen*)', Carr (1990) draws attention to the difficulties of the content of 'facts'. He comments (1990: 11): 'It used to be said that facts speak for themselves. This is, of course, untrue. The facts speak only when the historian calls on them: it is he who decides which facts to give the floor, and in what order or context.' Therefore it can be argued 'historical facts' are selective; for example, although I had breakfast this morning is a fact, it is unlikely to be entered into the Chronicle of World History. The status of a historical fact is subsequently likely to be determined by the ownership of history and the interpretation of its significance.

The issue of the ownership of the picture of the past is elaborated upon by Carr (1990: 11):

> History has been called an enormous jig-saw with a lot of missing parts. But the main trouble does not consist in the lacunae. Our picture of Greece in the fifth century BC is defective not primarily because so many of the bits have been accidentally lost, but because it is by and large, the picture formed by a tiny group of people in the city of Athens.

Such bias is also evident in the historical account of tourism, for example Towner (1996) observes that our understanding of the Grand Tour is mainly from the diaries of the wealthy, giving us an incomplete picture of the other people who have been in the Tour or the impressions of the communities that received these early tourists. Notable lacunae also exist in the history of tourism in most parts of the world, with the exception of Western Europe and to a lesser extent North America and Australia.

Connected to issues of ownership, and whose history is presented to us, a further issue is of the availability of materials from which historians can put together the past. Historians rely heavily on written evidence, including primary resources such as original documents and inscriptions (Jordanova, 2000). Yet the survival of documents is somewhat arbitrary and as both Carr (1990) and Towner (1996) suggest, the establishment of historical fact is limited by its ownership of who could produce such materials and interpret them. This reliance is likely to overemphasise the history of the wealthy and elites of society, those people that throughout history have had access to the educational resources to learn to write and read. There is also a lack of a feminine voice in history; consequently we know comparatively little about the role of women in tourism over time, either as travellers or as 'hosts' in tourism destinations.

Despite these limitations, the discipline of history has a vital role to play in aiding our understanding of tourism as Towner and Wall (1991: 72) emphasise: 'History is concerned with the dimension of time and attempts to understand social processes and institutions within this context. ... Fundamentally, history considers the transformation of things (people, places, institutions, ideas), through time, from one state into another.' Subsequently, it has a vital role in combination with geography, in helping us understand the transformation and development of tourism through time. The approach taken in this account of the history of tourism is to emphasise the processes occurring in places that can help us understand patterns of contemporary tourism.

Tourism and classical civilisation

While for many people living in economically developed countries, being a tourist is a regular if still special experience; the concept of a mass participation in tourism is a relatively recent phenomenon. However, the evidence of tourism can be traced back to the ancient civilisation of Greece. Although participation in tourism was constrained by the difficulties and dangers of travelling any distance, and also by a lack of financial resources to do so, there was evidence of travel by the Ancient Greeks for the purposes of oracles, festivals and game competitions. For example, the oracle of Delphi was highly significant in attracting people seeking advice about various matters of their lives, while many sick people travelled to Epidauras to be cured by the gods (Sharpley, 1999).

In 776 BC visitors from all over the Hellenic world attended the first Olympic Games at Olympia (Burrell, 1989). The games also possessed a strong religious dimension, with archaeological excavations at Olympia in 1876 providing evidence of the Greek gods playing an instrumental part in the victory of the athletes. More than 13,000 small to medium-sized bronze objects were unearthed, given as an offering for a thanks of victory. The site of an altar to Zeus was also found, where, apparently, athletes promised to obey the rules of the games (Burrell, ibid.).

In the Roman Empire, the next major civilisation of Europe, travel was facilitated by a sophisticated road system stretching 4,500 miles, the need to use only one currency, and a common language of Latin. The Roman Empire extended from Britain in the west to Armenia and Syria in the east, lasting for nearly 500 years in the west (c.31 BC–AD 476) and for 1,500 years in the east (c.31 BC–AD 1453), necessitating the building of an extensive road network to administer and maintain order. However, as Sharpley (1999) points out, long-distance travel for reasons other than trade and military service were uncommon.

As the city state of Rome grew, reaching a population of one million in Augustan times (Goudie and Viles, 1997), the desire of wealthy Romans to escape the heat in summer time led them to travel to the west coast of Italy, and the Bay of Naples developed as a fashionable tourist area. Subsequently, as Sharpley (1999) suggests, the Romans can be viewed as introducing the concept of tourism as a form of escapism from the home environment. Wealthier members of Roman society owned villages in the Alban and Sabine Hills around Rome and also in the coastal resorts of the Bay of Naples.

The extent of the tourism development in this area was to remain unmatched until the development of the French Riviera in the nineteenth century, nearly 2,000 years later (Davidson and Spearritt, 2000). An early form of resort hierarchy also emerged, with particular destinations having their own cultural environments, as Holloway (1998: 17) comments:

> Naples itself attracted the retired and intellectuals, Cumae became the resort of high fashion, Puteoli attracted the more staid tourist, while Baiae, which was both a spa town and a seaside resort, attracted the down-market tourist, becoming noted for its rowdiness, drunkenness and all-night singing.

Originally Baiae had been favoured as a place to go to bathe in hot springs in winter but as tourism developed in the summer it gained a reputation for moral laxity, with excessive drinking and nude bathing being commonplace, bringing in turn condemnation from some members of Roman society (Sharpley, 1999).

Think point

Historical evidence suggests that in Roman times resorts around the Bay of Naples catered for different types of travellers. How is the pattern of destinations catering for different types of tourists repeated in contemporary times?

A further point of consideration about tourism in Roman times was that similar to Ancient Greece, Roman society encompassed slavery. Hence, travel for the purpose of pleasure was restricted to a class who had both the available leisure time and disposable income to participate in it. It is not unrealistic to expect that this elite class of society would have been conspicuous by its ability to participate in tourism, a situation that existed until the nineteenth century, when the marked economic and social changes associated with the Industrial Revolution would eventually permit a wider social participation in tourism as is discussed later in the chapter.

Medieval tourism

Following the demise of the Roman Empire until the Renaissance period of the sixteenth and seventeenth centuries, travel became more difficult and limited. This was attributable to a lack of technological development of transport, a poor road infrastructure, and an absence of safety from aggression and robbery when travelling. Opportunities for travel were also restricted by the widespread poverty that existed during this time in Europe (Sharpley, 1999).

During this period of approximately 1,000 years c.500–1500 AD, often referred to as the 'Middle Ages', travel was arduous and mostly being undertaken out of a necessity to trade or for religious pilgrimage rather than for recreation. While pilgrimages were expected to be arduous, demonstrating religious devotion and acting as a form of penance, they often encompassed elements of pleasure as exemplified in Chaucer's *Canterbury Tales* (Urry, 1990). According to Davidson and Spearritt (2000) religious pilgrimages often included illicit souvenir hunting and sexual escapades, having some of the similar characteristics of the less endearing aspects of contemporary tourism. The debauchery associated with pilgrimages eventually led the Church to condemn this 'epidemic' of medieval tourism (Brendon, 1991). The three main destinations for pilgrimage were Jerusalem, Rome and Santiago de Compostella in north-west Spain, all attracting large numbers of pilgrims, with approximately 300,000 pilgrims visiting Rome in 1300 (Sharpley, 1999).

Throughout the Middle Ages, religion played an important part in folk culture, and the celebrating of 'holy days', from which the word 'holiday' eventually developed, presented opportunities for a change from the typical employment in agriculture and cottage industries. As opportunities for travel were limited, 'popular culture' was locally based, being defined by Burke (1978: 1) as: 'the culture of the non-elite, the "subordinate classes" as Gramsci called them.' As the main threat to security when travelling came from highwaymen who robbed travellers, travel opportunities were subsequently usually restricted to the well-guarded royalty and the court circle, and a handful of other wealthy citizens (Holloway, 1998). However, it would be misleading to assume that social class always differentiated leisure and other types of enjoyment during this time. Burke (1978) comments that there is plenty of historical evidence to support the participation of the upper class in popular culture, notably at festival time.

The Grand Tour

The catalyst for the movement of a larger number of the wealthy beyond national boundaries was associated with the Renaissance, a period of revived interest during the sixteenth and seventeenth centuries in the classical civilisations of Rome and Athens. Historical records highlight the influence of the British aristocracy in travel at this time, which is attributable to the encouragement by Elizabeth I of young men seeking positions at court to travel to the mainland of Europe to study classical culture (Holloway, 1998). As Brendon (1991: 10) comments: 'It [the Grand Tour] has begun, during Queen Elizabeth's reign, as a refined form of education, a school to "finish" patricians by giving them first hand experience of classical lands.' Others in the upper class of society copied this practice, and travel to Italy for the purposes of teachings in classical civilisation eventually became a part of a gentleman's education. This upper-class travel dating from the Renaissance period became known as the 'Grand Tour', as it expanded beyond Italy into a circuit of Western Europe. The monuments and history of Roman civilisation remain an attraction for tourists, just as they did for participants of the Grand Tour, as is depicted in Figure 1.5.

In its early period the main theme of the Grand Tour was educational as Inglis (2000: 17) comments: 'The maxim that a tourist should be improved and educated by the Tour was inscribed in the practice from the start.' Leed (1991) also emphasises the importance of education as a focus for the Grand Tour, referring to the *peregrinatio academia*, or the 'journeyman's scholar year' as an essential component of a gentleman's upbringing. Notably from the perspective of reasons to travel, the Grand Tour introduced the theme of education as a new dimension to travel and turned the purpose of touring into a civilising and cultivating process.

The Grand Tour also gave rise to a new profession of the 'travelling tutor' who was to:

> watch over the morals of the travelling nobleman, act as a guide, see to accommodations, introduce him to the arts, books, and learned men, and gauge his process in the courtly and literary skills that were increasingly the legitimation of nobility.
>
> (Leed, 1991: 185)

One of the most famous travelling tutors was Adam Smith, the eminent economist whose seminal work, *An Inquiry Into the Nature and Causes of the Wealth of Nations* published in 1776, has had a major influence upon economic thought as is discussed in Chapter 4.

Figure 1.5 Remnants of the classical civilisation of Ancient Rome.

An essential component of this educational process was the keeping of a journal by the traveller, 'as a kind of memorial to his investiture in the world and as a way of halting the erosions of memory by time' (Gill, 1967: 53). It is from the analysis of the content of these diaries that the recounting of events of the Grand Tour are largely based. Consequently, this gives an often Anglo-centric bias towards the accounts of the Grand Tour (Towner, 1996). Also, while there exist written accounts of the experiences of the tourists, there is an absence of accounts of the thoughts and impacts of the tours upon the communities and residents in destination societies.

Although the original focus may have been upon Italy, the Grand Tour spread geographically to incorporate other major cultural centres of Europe, leading Inglis (2000) to suggest that the word 'tourism' can be dated to this period. This spatial diffusion of travel led Towner (1996) to suggest that it is uncertain how to best interpret the Grand Tour, i.e. either as where the aristocracy travelled to, or as a defined tourist circuit of Western Europe. If the former approach were adopted than the geographic extent of the Grand Tour would be extended to include Portugal, Spain, Greece, the Near East, Russia and Scandinavia (Towner, ibid).

As the eighteenth century progressed, more people could afford to participate in the Grand Tour as a consequence of an expansion in mercantile trade and subsequent increased wealth. Another significant factor in facilitating travel was political stability in Europe, as a consequence of the signing of the Peace of Paris Treaty in 1763 by France, Spain and Britain, which ended intermittent hostilities between the countries that had lasted for most of the preceding century (Inglis, 2000). In the two years after the signing of the Treaty of Paris it is estimated that 40,000 English travellers passed through the port of Calais bound for France (Brendon, 1991).

With the increase in the numbers of participants in the Grand Tour came a widening of the variety of types of people and a diversification of themes. Progressively through the eighteenth century, the middle classes began to form the majority of tourists rather than the aristocracy, which led to a significant rise in family and women travellers (Towner, 1996). By the latter half of the eighteenth century, Brendon comments that the Grand Tour had become little more than a: 'refined form of pleasure, a Continental jaunt for those who wanted to keep up with the milords' (1991: 10).

Think point

The Grand Tour was established by the aristocracy in the seventeenth century and by the end of the eighteenth century had become, in the words of Brendon (1991: 10) a: 'refined form of pleasure, a Continental jaunt for those who wanted to keep up with the milords.' How do role models and fashion influence where people travel today?

Travel for health reasons added a new dimension to the tour, with the south of France becoming a focal point to deal with the ill effects of the climate of the north of Europe. Montpelier in France developed as a health resort to offset the effects of consumption in the seventeenth century (Towner, ibid.) and other destinations on the French Riviera, notably Nice, emerged as places to travel to for health cures in the eighteenth century

(Nash, 1979). By the early nineteenth century there were accounts of the town filling with visitors from England escaping the winter, although as Nash (ibid.) points out, by this stage healthy tourists probably outnumbered the infirm.

Travel for the purposes of amusement and pleasure, to enjoy the cultures and social life of cities like Paris, Venice and Florence, also progressively became part of the Tour, and by the end of the eighteenth century this custom had become an institutionalised component of it (Holloway, 1998). From the mid-eighteenth century, a much greater emphasis also began to be placed upon the viewing of nature, emphasising its spiritual and romantic qualities (Towner, 1996). Typically this entailed the viewing of 'wildscape', such as mountain areas, where there was little human alteration of the landscape.

Think point

As the Grand Tour progressed during the seventeenth and eighteenth centuries a number of themes of travel are evident. These include education, health, enjoyment of culture and visiting nature. How relevant are these themes to tourism today?

This wish to view wildscape marked a shift from what were regarded as desirable landscapes. The previous landscapes of fashion were those of the European low countries, i.e. Belgium and Holland, because they illustrated the human ability to control and dominate nature and provide agriculturally productive terrain. By contrast, the desire to view 'wildscape' was marked by a preference for the raw power of nature, as manifested in mountains, gorges, waterfalls and forests. Until this time barren and mountainous landscapes had been largely detested and even feared by the majority of the population. For example, Smout (1990) points out that up to the eighteenth century, the environment of the Scottish Highlands in Britain had been largely seen as an inhospitable place to think of visiting and any use of its environment had strictly been in utilitarian terms such as for the cutting of wood and mining of ores. Urry (1995: 213) adds: 'It is only in the last century [the nineteenth century] that a traveller passing through the Alps would have the carriage blinds lowered to make sure they were not unduly offended by the site.'

The increase in the popularity of wilder landscapes was associated with the development of the 'Romantic movement', which stressed the feelings of emotion, joy, freedom and beauty that could be gained through visitation to 'untamed' landscapes. The Romantics were a collective movement of European literary, artistic and musical figures, including Rousseau, Coleridge, Wordsworth, Chopin, Goethe, Walter Scott, Hugo, Liszt and Brahms, who highlighted the importance of emotional experiences and feelings about the natural and supernatural world. The movement was in part a reaction to the scientific thinking of the Enlightenment period, and also to the creeping urbanisation of Britain and Western Europe associated with the Industrial Revolution. The Romantics subsequently represented a form of political opposition to what they perceived to be a loss of community, associated with the migration of people to urban areas from rural environments. The defining of attractive landscapes as wild areas has had a major influence on patterns of contemporary tourism. The influence of

Romanticism led to a newfound appreciation of not only mountain areas, but also the coastline, the two most important areas for recreational tourism.

A further development in tourism during the time of the Grand Tour was an increased popularity in the spa towns of Europe, the first time since the decline of the Roman Empire. Based upon an association between the medicinal qualities of taking the spa waters and good health, spas represented an early type of health tourism. During the eighteenth century spa towns such as Bath in England, Vichy in France and Baden-Baden in Germany reached the height of their popularity. While at its most basic a spa consists of a pump room for water and a bath, this was often followed by the development of a range of other activities such as assembly rooms, ballrooms, gambling casinos and high-class brothels (Hobsbawm, 1975), which Walton (2002) suggests formed their real 'raison d'être'. The popularity of spa towns also involved a cultural dimension, their cosmopolitan clientele allowing people to bid for status through fashion and achievement, rather than hereditary lineage or known office. However, by the end of the eighteenth century spa towns were turning primarily into residential and commercial centres, as coastal areas grew in popularity for tourism.

The eventual demise of the Grand Tour in the early nineteenth century is usually attributed to the outbreak of the Napoleonic Wars. However, in the latter two decades of the eighteenth century major economic and social changes were taking place, notably in Britain, which were also to occur in other European countries and North America during the nineteenth century. These changes were ultimately to lead to the wider social participation in tourism with which we are familiar today. This period of marked economic and social change is referred to as the 'Industrial Revolution', regarded by the eminent historian Eric Hobsbawm (1962) as probably the most important event in world history.

The Industrial Revolution: laying the roots for contemporary tourism

The effects of the Industrial Revolution upon society were, as Hobsbawm (1962) suggests, enormous. These included economic, social, political, technological and cultural changes, the interaction and amalgam of which, besides radically changing society, help to explain patterns of contemporary tourism. As the name suggests, the Industrial Revolution is significant for marking a period of change from an agricultural-based economy to an industrial one. Its origins lie in the mechanisation of cotton and wool production in the north of England in the last two decades of the eighteenth century.

Although the Industrial Revolution began in Britain, other countries have followed similar patterns of change. Commenting on the take-off of different countries towards industrialisation Rostow (1971: 9) observes:

> one can approximately allocate the take-off of Britain to the two decades after 1783; France and the United States to the several decades preceding 1860; Germany, the third quarter of the nineteenth century; Japan, the fourth quarter of the nineteenth century; Russia and Canada the quarter-century or so preceding 1914; while during the 1950's India and China have, in different ways, launched their respective take-offs.

In terms of explaining how the Industrial Revolution influenced the development of mass participation in tourism, certain key themes are evident: urbanisation and increased economic production; the technological advancement of transport and the emergence of a tourism industry; and the development of the seaside as a spatial area for 'mass tourism'.

Urbanisation and increased economic production

A key change in all countries undergoing the Industrial Revolution was the movement of the workforce from rural areas to urban centres and the development of a new social structure, including the new entrepreneurial class called the 'bourgeoisie' and industrial working class or 'proletariat'. As a prototype of what other industrialising countries were to follow to varying degrees of similarity, Britain became progressively urbanised during the nineteenth century. Not only was the pace of urbanisation fast but it also resulted in a significant change in lifestyle. In 1801 there were no towns in Britain outside London with more than 100,000 inhabitants, but by 1831 there were seven, accounting collectively for one-sixth of the population. In reference to the significance of this one-sixth in establishing themselves as the avant-garde of a changing society, Thompson (1988: 28) comments: '[it was] here that new habits and lifestyles were evolved which ultimately percolated through to the rest of the country, and here that familiar and accustomed patterns of behaviour were most strongly challenged'. By 1901, 80 per cent of the population of England and Wales lived in towns, compared to a mere 20 per cent in 1801 (Urry, 1990). Choosing 1851 as a vignette of pivotal changes in British society, Clarke and Critcher (1985) comment that this was the first time that the majority of the population lived in urban areas, and that 30 per cent of the workforce was now employed in industry compared to the 20 per cent working in agriculture.

Similar urban development in the latter half of the nineteenth century was also taking place in other European countries: America and Australia. For example, from having populations of fewer than 50,000 in 1850, the populations of San Francisco in the US and Melbourne in Australia had risen to 364,000 and 743,000 respectively by 1891 (Soane, 1993). The total number of inhabitants in towns of over 100,000 people in North America and Europe increased from 14,800,000 in 1850 to 80,800,000 by 1913 (Soane, ibid).

The rapid growth of towns and cities and lack of town-planning gave rise to a number of social issues, not least the living conditions of the proletariat. The town of Manchester in the north of England acquired a reputation as the 'shock' city of the Industrial Revolution, with the poor social conditions recorded by Friedrich Engels, the co-author with Karl Marx of the Communist Party manifesto. Graphically commenting on the conditions of the proletariat in Manchester, Engels (1845: 100) states:

> If we briefly formulate the result of our wanderings, we must admit that 350000 working people of Manchester and its environs live, almost all of them, in wretched, damp, filthy cottages, that the streets which surround them are usually in the most miserable and filthy condition, laid out without the slightest reference to ventilation, with reference solely to the profit secured by the contractor. In a word, we must confess that in the working men's dwellings of Manchester, no cleanliness, no convenience, and consequently no comfortable family life is

possible; that in such dwellings only a physically degenerate race, robbed of all humanity, degraded, reduced morally and physically to bestiality, could feel comfortable at home.

Urbanisation caused the separation of people from nature and the land for the first time in human history. The workforce was also required to work in a manner that was suited to the needs of industry and factories rather than the natural rhythms of the seasons. Factory work required a regular unbroken daily routine, as life became structured around the need to keep industrial production functioning, with men, women and children often working a six-day and 70-hour week (Clarke and Critcher, 1985). Employers often complained about the laziness of their employees, their tendency to work until they had earned a traditional week's wage, and then to stop (Hobsbawm, 1962). The chief method to tackle this was to pay the workforce so little it would need to work longer. However, whether as a consequence of a newfound philanthropy among factory owners, or a realisation that a healthier workforce would be a more productive one, from 1850 onwards there was a general decrease in working hours so that by the 1870s a nine-hour day was the norm (Towner, 1996).

The Industrial Revolution also led to a clearer differentiation of periods of time and activities between work and leisure than had previously existed. In their analysis of the changing structure and practices that resulted from the Industrial Revolution, Clarke and Critcher (1985) comment that previously there had been no clear demarcation between work and leisure. For example, drinking while at work and breaking-off from work to attend to domestic affairs, which were not unfamiliar practices in agrarian societies, were aspects of work life that factory owners could not tolerate. Similarly, the interaction of people while at work, for example chatting or playing, were features of agricultural societies that were not compatible with factory production. The folk culture of village communities was also lost as people were forced to live in poor conditions, similar to the ones described by Engels, creating social tensions that had never been experienced before (Soane, 1993). This pattern of separate spatial and time zones for work and leisure is reflected in contemporary tourism, as we take defined periods of time away from work, and travel away from our home environment to other places and destinations.

Think point

How do you think living in a city or town environment influences the desire to participate in tourism? Consider the kinds of environments people wish to travel to and compare them to the ones they come from.

An outcome of major significance of the Industrial Revolution was the level of productivity that was established in the economy. Nash (1996) comments that a high level of productivity is the key to the establishment of a leisure class able to participate in tourism. As the Industrial Revolution progressed the national income per head in England and Wales quadrupled during the nineteenth century (Urry, 1990).

Consequently, people began to have disposable income, i.e. extra income left over after spending on essential items such as housing and food, to spend on leisure activities. A consequence of the availability of income to spend on leisure was the development of the tourism industry as is discussed in the next section of this chapter.

Besides having money to spend on tourism a further requirement is time free from employment. This can be facilitated by legislation from government, as was the case in the UK with the passing of the Bank Holidays Acts of 1871 and 1875, which provided a four-day statutory holiday (Towner, 1996). A watershed was subsequently passed in terms of the recognition of balancing work time with statutory leisure time. Certainly political pressures, including the heightened profile of the Trade Union movement, the founding of the socialist Labour Party, the social fallout from the First World War and a consequent demand for rights for workers, were contributory factors to the passing of the 1938 Holiday with Pay Act in the UK. Similarly in France, government legislation was passed in 1936, making 12 days of paid vacation mandatory in all enterprises (Dumazedier, 1967). The significance of these acts is that it marks recognition of holidays as being beneficial for individuals and society.

Think point

How do levels of income and holiday time influence patterns of contemporary tourism? Is there specific government legislation in your own country that permits workers paid periods of leave from work?

A transport revolution and entrepreneurial activity

A further essential element of contemporary tourism is reliance upon fast and efficient transport. The Industrial Revolution was characterised by a technological advancement in travel, notably the invention of the steam engine by James Watt in 1784, which led to the development of the railway and steamship. Until the nineteenth century, travel was dependent, as it always had been, upon horse and wind power (Brendon, 1991), often being arduous, for example a journey of approximately 640 kilometres between London and Edinburgh took ten days by horse and carriage (Holloway, 1998). A revolution in travel was marked by the opening in 1830 of the first railway line in the world to carry passengers. It ran between Liverpool and Manchester in England and during that decade speeds of up to 60 miles (100 kilometres) per hour became possible on the railways (Hobsbawm, 1962). The steam engine was also utilised to increase speeds and travel on the seas with the development and use of steamships. The first regular cross-Channel service between England and France began in 1820 after the end of the Napoleonic wars, making travel easier and reliable, and it is estimated that by 1840 as many as 100,000 people per annum were using it (Brendon, 1991). In 1841 the first steamship crossed the Atlantic from Britain to America.

The development of the railway network in the nineteenth century had a marked significance for both society and tourism. The importance of the railways is underlined by Thompson (1988: 46) who comments: 'Symbolically and literally the railway lay somewhere near the centre of that society [Victorian].' Besides being instrumental to the Industrial Revolution, by moving coal from the coalfields to the factories to power the steam engines to drive the machinery, the railways were a major force in eroding localism and removing barriers to mobility. They brought together different regions of the country and eventually different countries within Europe. During the nineteenth century, the railways were also instrumental in developing tourism in the US and Australia. As Hobsbawm (1962: 60) points out: 'No innovation of the Industrial Revolution has fired the imagination as much as the railway.' The construction of the railway system was vital to the development of coastal and mountain areas for tourism and also organised tours as described in Box 1.1.

Thompson (1988) makes an interesting observation of an early differentiation of themes and markets in the operations of Thomas Cook. Commenting on Cook's organised tours from the provinces by rail to the Great Exhibition at Crystal Palace in London in 1851, he notes that they were based upon a philanthropic rationale aimed specifically at the 'respectable' working classes, with the aim of exposing them to the educational message of the exhibition. The themes of that message included material, cultural and moral progress, alongside peaceful international cooperation.

The individual contribution of Cook to bringing visitors to the Great Exhibition was notable; of the total six million visitors, he brought 165,000 of them (Brendon, 1991) or nearly 3 per cent. In contrast, Thompson (1988) comments that foreign travel

---

Box 1.1 A revolution in travel: Thomas Cook and the railways

By the 1840s the potential of the railways for tourism was already being realised by Thomas Cook, a wood turner by trade, secretary of the Midland Temperance Association in his spare time, and a believer that alcohol was the root of the majority of evils in Victorian Britain (Brendon, 1991). The seminal event in the beginning of the use of the railways for recreational tourism was the organisation of a trip for 570 temperance workers from Leicester to a temperance rally near Loughborough by train in 1841. This trip demonstrated the potential demand for group travel, while Cook also realised the potential of his own power as a bargaining agent to capture reduced group prices with the railways and other suppliers of travel services (Thompson, 1988). By 1845 he was arranging similar excursions on a full commercial basis using chartered trains. Cook's efforts represented the beginnings of the development of the tourism industry, while for Brendon (1991: 17): 'Thomas Cook's achievement was to associate himself with the spirit of the age and to foster its most thrilling development.' The efforts of Cook can certainly be equated with a revolution in travel, simplifying, popularising and cheapening travel, to bring it within the reach of the working classes. In this sense he was instrumental, in combination with the railway, in helping transform society from a largely static one in which the poorer stayed at home, into a more mobile and fluid one.

organised by Cook was targeted specifically at the middle classes and had a much stronger commercial orientation. He comments (1988: 262) that trips to: 'Paris, and the French, Swiss and Italian resorts, from the later 1850s and early 1860s, were middle-class affairs with middle-class fares.'

An advertisement for Cook's tours to Europe dated 1904 is shown in Figure 1.6. The influence of the steam engine is evident with the use of both the railways and steamships, indicated by the reference to travel via Dover to Calais and Folkestone to Boulogne, forming an essential part of the tour. From the dress and appearance of the tourists in the poster it is also evident that the target market for these tours was the upper-middle class and the bourgeoisie; the aristocracy would not have considered taking part in a package tour.

By the 1860s, Cook had already developed tours to Europe and America, and in 1869 offered the first escorted tour to the Holy Land (Boorstin, 1961). In his first nine years of business, he handled more than one million customers (Eadington and Smith, 1992). By the beginning of the twentieth century, Cook and Son had started to make arrangements for travellers all around the world, but the bulk of their business remained in Europe (Wigg, 1996). However, not everyone looked favourably upon Cook's efforts to democratise travel and encourage a wider participation in tourism. Assorted labels were given to the groups organised by Cook, including 'Cook's Circus'; 'Cook's Hordes'; 'Cook's Vandals'; while his visitors to Switzerland were labelled as a 'low vulgar mob' and a 'swarm of intrusive insects' (Brendon, 1991). Hopefully, appropriate retorts were given to those who did the labelling, but such views display a culturally elitist attitude to travel still witnessed today.

Think point

Cook's organised groups of tourists to Europe were termed 'Cook's Circus'; 'Cook's Hordes'; 'Cook's Vandals' by some aristocratic independent travellers. Is there evidence today of such kind of criticism of tourists and cultural elitism?

The destinations available for recreation began to increase and diversify as the railways made possible regular and safe journeys for the first time in history. As Hobsbawm comments (1975: 240): 'Industrial capitalism produced two novel forms of pleasure travel: tourism and summer holidays for the bourgeoisie and mechanised day trips for the masses in the countries such as Britain.' In terms of establishing the pattern of contemporary tourism, the railways were instrumental in giving access from urban areas for the masses to the coastline.

The developing seaside and mass participation in tourism

The first geographical area to become a focus for mass participation in tourism was the coast. Although Brendon (1991) suggests that the popularity of the seaside was encouraged by royal patronage in the eighteenth century, in the manifestation of the mentally unstable British monarch George III who hoped to find sanity through drinking salt

Figure 1.6 Advertisement for Cook's tours (1904).

water, this perspective ignores the existence of a pre-existing sea-bathing culture in other European countries. Towner (1996) draws attention to the use of the coast in peasant folk culture in the Baltic, North Sea and the Mediterranean, with a peasant sea-bathing culture existing in regions as diverse as the north of England and the Basque region of Spain, before its patronage by upper and middle classes. Towner (ibid.) suggests that the tendency to overlook this fact demonstrates how those that dominate the historical record are assumed to be innovators of custom.

The development of coastal areas was encouraged by changing landscape tastes of Romanticism during the eighteenth century, referred to earlier. The impact of urbanisation was also influential in encouraging a seaside culture as Towner (1996: 170) comments: 'Grafted onto the growing taste for coastal areas was the influence of health awareness and desire to escape from the effects of rapid urbanisation.' The historical records of the use of the coast place an emphasis on its use initially for health rather than pleasure. The desire for 'escape' is also stressed by Soane (1993), who suggests that coastal areas provided space to escape to in response to nineteenth century industrialisation, as the poor social conditions of the urban landscape encouraged the middle classes to develop anti-urban values.

During the eighteenth century coastal resorts began to rival spa towns as fashionable places for the growing middle classes in Europe and America to visit. However, Towner (1996) points out that the demand for the seaside varied internationally, with a sea-bathing culture first appearing in Britain in the 1730s, followed by France in the 1780s, Germany in the 1790s and Spain in the 1830s. He suggests that this pattern can be attributed to variations in wealth and leisure but also involves a cultural dimension of the type of landscape that was viewed as being favourable within different countries.

The ability to 'escape' to coastal areas was determined by a mix of economic and social factors. These included possessing the 'cultural capital' of knowing the qualities of the coast; having the time to visit; and also having the money to do so. As Soane (1993) points out, in the early nineteenth century only the wealthy could afford to stay at the coast for long periods of time, and the development of coastal residential resorts offered an exceptional degree of privacy to the new elite, away from the industrial urban centres and the proletarian masses.

In terms of developing a popular seaside culture, it was particularly the development of the railway network from the cities to the coast, which permitted a middle- and working-class holiday boom during the late nineteenth century and early twentieth century. Villages and towns on the coastlines of industrialising centres were transformed with promenades and piers, providing profits from previously economically redundant areas of cliffs and bays. The coast seemed to exercise an allure that eventually permeated all the social classes; Towner (1996: 212) suggests that the seaside represented a special place in many people's lives, which he refers to as the 'geography of hope'. Certainly, the distinctive natural landscape of sea, sky, cliffs and beach, and built resort landscapes of promenades and piers, provided a distinctive sense of place away from the ordinary and the routine.

While from the 1870s it became normal for the middle classes to take a holiday of two to three weeks at a seaside resort (Soane, 1993), in England the development of the working-class seaside resort did not become of major significance until the 1880s

(Hobsbawm, 1975). The railways played a major part in this development, moving thousands of working-class people from the cities to the coast, for example in 1858 it was estimated that 200,000 people left Manchester in England for the coast during the religious holiday of Whit week (Urry, 1990).

The combination of health, natural and created attractions, and the use of the railways for transport helped make coastal areas popular as resorts, as is shown in Figure 1.7 – a travel poster from the first decade of the twentieth century advertising the British resort of Southend and Westcliff-on-Sea. The poster emphasises a healthy and tranquil atmosphere, alongside a constructed 'pleasure periphery' of a human transformed landscape, including a promenade and pier. In reality, by 1909, the date of the poster, the chance of actually experiencing the tranquil kind of scene depicted was remote. Southend's population had grown from under 3,000 in 1871 to 63,000 by 1911 (Wigg, 1996), as it had become a very popular resort for day trippers from London, subsequently attracting a variety of types of workers needed to service the growing tourism industry.

Similar to Britain, the new middle classes and bourgeoisie were instrumental to the development of coastal areas in other industrialising countries in the latter half of the nineteenth century. The well-ordered appearance and attractive settings of the coast began to be appreciated in a variety of cultures and by increasing numbers of visitors as railway networks expanded into the remote parts of industrialising countries. Major areas of coastal development in the last quarter of the nineteenth century and early twentieth century were the French and Italian Rivieras from Canne to San Remo, and the southern Californian coast from Santa Barbara to San Diego (Soane, 1993). Similar to the growth of the British seaside resorts like Southend, as they attracted more visitors, in turn, they attracted more workers and permanent residents, consequently growing in size. Typically, employment in banks and other service industries was created as coastal towns took on a growth dynamic of their own.

However, it was not solely coastal areas that developed in the nineteenth century as popular areas for tourism. A further important development was the popularisation of mountain areas for winter tourism, as is explained in Box 1.2.

The force of the economic and social changes that originated during the Industrial Revolution and moved society from a mainly static one to a mobile one, as summarised in Figure 1.8, continued to make itself felt in the twentieth century.

While the trend towards mass participation in domestic tourism was maintained for the first part of the twentieth century, economic recession in the 1930s combined with two world wars in the first half of the century greatly restricted the opportunities for the growth of international tourism. However, a second phase of mass tourism occurred in the 1950s after the Second World War that, this time, had an international perspective, as is discussed in Box 1.3.

Extending the peripheries and changing markets

The changing economic, social and technological societies that have created mass tourism continue to drive its demand. Just as the numbers of people participating in tourism increased continuously through the twentieth century, the peripheries of tourism

Figure 1.7 Advertisement for Southend, England (1909).

Box 1.2 The development of mountain areas for winter tourism

Besides the popularisation of coastal areas in the nineteenth century, another landscape to be used for tourism was mountain areas, particularly for the development of winter sports. Towards the end of the nineteenth century skiing began to develop as an international activity in the European Alps. Resorts such as Davos and St Moritz were already familiar to wealthier travellers, for the purpose of health tourism, based on the 'cold cure', which was a popular treatment. By the end of the nineteenth century, the popularity of the European Alps had increased with the expansion of the railroads into Alpine valleys (Barker, 1982). St Moritz in Switzerland, St Gervais in France and Badgastein and Bad Ischl in Austria were established as health spas by the end of the nineteenth century.

However, in the 1890s a new type of traveller appeared in the Alps more intent upon hedonism than recuperation, with winter sports, including ice-skating and skiing, becoming fashionable and popular. The mountains had become increasingly popular to upper-class Victorians from the beginning of the nineteenth century, as an escape from the growing urbanised areas of the Industrial Revolution, besides for health purposes. D'Auvergne (1910: 289) states: 'The tide of fashion has in fact been largely diverted of recent years from Nice and Cairo to these snow bound wildernesses.' An essential part of this fashion was the development of skiing, notably popularised by Sir Arthur Conan Doyle in his crossing by ski from Davos to St Moritz in Switzerland. The Alps now became the playground for Europe's elite.

By the 1920s, a significant winter season in the Alps based upon destinations accessible by the railways had already developed. An indication of the growing importance of the winter market in the Alps is demonstrated by statistics from the Austrian Tyrol. In 1924, 14 per cent of the overnight tourist stays in the Tyrol occurred in the winter season, which by 1933 had increased to 44 per cent of the annual total of overnight stays (Barker, 1982). However, the development of skiing as a recreation was not confined to Europe. In Elyne's (1942) account of skiing in the Australian Alps she says that the first skiing there took place in 1897. Good and Grenier (1994) give an earlier date, stating that the first ski club in the region was established in the 1860s gold rush. From these origins an industry has developed that now serves millions of skiers per annum visiting different mountain areas around the world. For example, three million foreign skiers per annum are attracted to Switzerland and Austria, and two million to France and Italy (Mintel, 1996).

What was originally an elitist activity has progressively become one of mass participation. Governments eager to aid regional economic development have supported this demand. The economic potential of skiing in Alpine areas had already been noted by the beginning of the twentieth century. D'Auvergne (1910: 280) comments:

> Winter sport! the Swiss delightedly awakened to the commercial possibilities of snow and ice. The canton Grisons or Graubunden – the largest in Switzerland – was the first to find foreign gold beneath the snowdrifts. . . . Naturally the rest

of Switzerland is on the alert and eager to share the good fortune of the largest canton. . . . Chalets were transformed into hotels, brand new hotels were run up not always to the delight of the aesthetic traveller.

The post-war French government also acted upon the realisation that skiing could aid regional economic development. Lewis and Wild (1995) explain how the purpose-built ski resorts – or 'ski factories' as Lewis and Wild refer to them, because of their emphasis on accommodating large numbers of skiers and construction from glass, concrete and steel – were developed in the late 1950s to aid regional development in France. The demand of the emerging 'mass leisure class', and the opportunities to aid regional economical and social development, drove the transformation of mountain landscapes in the Alps.

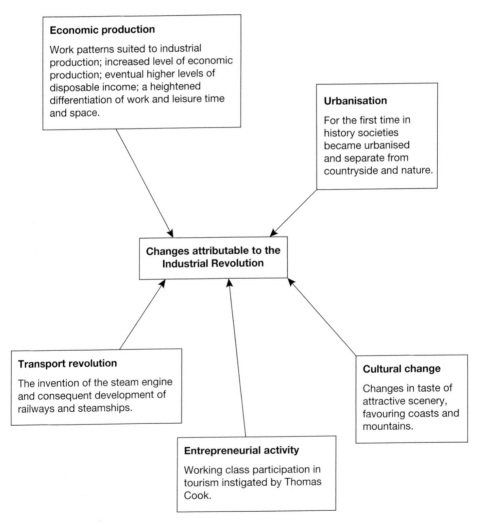

Figure 1.8 Changing from a static to a mobile society.

Box 1.3 The second wave of mass tourism – an international perspective

The ability of the working class to participate in tourism to the coasts of their own countries from the last decade of the nineteenth century marked a democratisation and revolution in travel. A second revolution of mass tourism began in the 1950s, based upon working-class participation in 'international tourism'. This trend emerged first in Britain, though why this should be the case is uncertain. While a range of economic and social factors, combined with entrepreneurial activity, coincided to explain this growth, similar social conditions existed in other countries. It may be that the more rigid class structure of Britain compared to many other European countries and its geographical isolation as an island, making travel to other countries more difficult, also played a part.

A range of economic and social factors combined to create the conditions for the participation of increasing numbers of people in travel overseas. The Holidays with Pay Act (1938) was introduced; the Education Act of 1944 raised the school-leaving age to 15 in 1947 and resulted in an expansion of grammar schools raising the aspirations of a whole new stratum of society; demand for labour created greater bargaining power for workers after the war pushing up wages; the increasing audience for television also influenced the desire to travel; and, critically, average weekly earnings rose by 34 per cent while retail prices increased by only 15 per cent between 1955 and 1960.

Also critical to this expansion of mass travel was the entrepreneurial activity of the Russian émigré to Britain, Vladimir Raitz, who founded the Horizon Travel Company. Just as Thomas Cook played a major role in widening the social participation in travel and tourism in the nineteenth century, the efforts of Raitz were fundamental in establishing a blueprint for mass international tourism. Putting together a package of air transport and accommodation, using the services of an air broker to charter aircraft rather than buying seats on a scheduled flight, Raitz accompanied 32 passengers in May 1950 on a converted Dakota with a top speed of 170 miles per hour from London to Calvi in Corsica.

Raitz had put together the first 'package tour' based upon air travel. Although Thomas Cook and other operators such as Henry Lunn offered group travel to other countries, they were based upon travel by rail and ship. The use of aircraft was fundamental to expanding the peripheries of tourism from Western Europe. This trend became even more noticeable with the technological development of the jet engine in the 1950s. The 'bulk' buying of seats on aircraft; hotel rooms; the provision of meals; and services of a company representative in the destination were to provide an attractive formula in the tourism market. Not least because foreign travel became competitively priced compared to domestic holidays. There was also the added attraction of environmental factors, notably almost guaranteed sunshine and a warm sea.

Today, the buying of similar packages based upon a formula of air travel and accommodation is familiar in many countries. Major tour operators, such as Touristik Union International, have become major transnational companies listed on stock markets with public shareholders. In a period of just 50 years the industry has matured notably from its origins, a reflection of the economic and financial importance of tourism, and the important role of tourism in many societies.

After: Bray and Raitz (2001)

have also been extended progressively, as is shown in Figure 1.9. There are few places in the world that remain untouched by tourism and with the development of space tourism, the periphery is now set to be expanded beyond the boundaries of the planet.

A notable evolution in the market for tourism manifested itself in the later 1980s with a demand for a diversity of destinations and new types of experience through tourism. A new generation of tourists, more confident and familiar with travel, and more independently minded than their parents, drove these changes. These tourists are referred to by Poon (1993) as 'new tourists' who are less predictable and homogeneous than the 'old tourists', characterised by being less interested in package holidays and group travel, instead wanting to emphasise their individuality and have control of their own experiences. Poon (ibid.) suggests that in place of the vacation representing an escape from work and home, new tourists go on holiday to experience something different, consequently integrating travel and vacations into their lifestyle.

Alongside the emergence of new tourists, there has also been a diversification in the types of tourism, with terms such as 'eco', 'cultural', 'alternative' and 'adventure' becoming part of the tourism lexicon. Besides being representative of a maturing of the tourism market, it is also important to understand these developments within the context

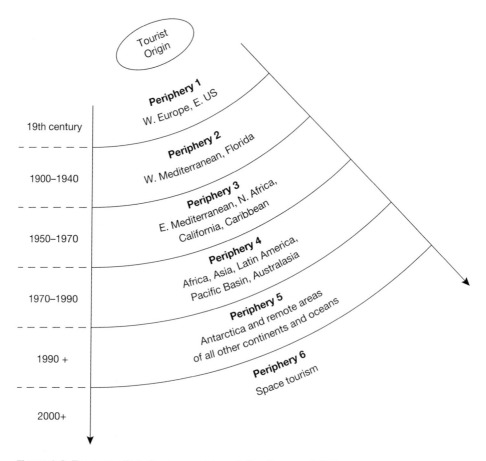

Figure 1.9 The expanding pleasure periphery (after Prosser, 1999).

of changes in the home society of tourists. Just as the popularisation of wildscape as part of the Grand Tour was attributable to Romanticism and urbanisation, similarly it is difficult to disassociate the emergence of ecotourism and cultural tourism. which emphasise the 'natural' and 'authentic', as being independent of a perceived disassociation from nature and pre-industrial culture by post-industrial societies.

Similarly, just as the entrepreneurial activity of Cook and Raitz was essential to a mass participation in travel, the efforts of new entrepreneurs in developing low-cost air travel have been instrumental in the lowering of the prices of air travel, diversifying the numbers of destinations, and increasing the opportunities for independent travel. The rapid progress in information technology in the last decade of the twentieth century, bringing into the home the possibility of making online airline and hotel reservations, has also had a great impact on the diversification of travel. As the WTO (2003c: 6) comment: '"Do-it-yourself"' is becoming more and more common, particularly for the mature and experienced travellers, vigorously stimulated by the possibilities offered by low-cost airlines and the Internet.'

Summary

- Tourism does not occur by chance, it is a product of changing economic and social factors. The major tourism-generating countries of the world are those with highly developed economies. Although the growth in participation in tourism has been dramatic during the last half-century, the majority of the world's population are excluded from participation, typically because of poverty. Similarly, climate is an influential factor in determining the destination areas to which recreational tourists will travel. The most popular tourism destination area in the world is the Mediterranean.

- One way of thinking about tourism that incorporates much of its complexity is as a system. This permits tourism to be thought of as being dynamic, open to various influences, and operating on different spatial scales. It is important to remember that within the system exists reciprocity. Subsequently, tourism brings changes not only to the societies tourists visit, but also to the ones they come from.

- Destinations that we view as desirable to visit are highly influenced by cultural perceptions. The two most popular areas for recreational tourism, coastlines and mountain areas, were popularised by the Romantic movement. This movement was also associated with the pivotal event in history in terms of influencing patterns of contemporary tourism, the Industrial Revolution. Beginning at the end of the eighteenth century, the populations of many countries underwent a radical change in lifestyle as a consequence of industrialisation. This included a move to living in urban towns in place of rural villages and working in factories instead of in agriculture and cottage industries. Critically, the Industrial Revolution created a level of economic productivity that permitted a mass participation in tourism. The technological advancement in transport, notably the railways, was also a key contributory factor to permitting a working-class participation in tourism, and in aiding society to move from being a largely static one to a mobile one.

■ While the domestic coastline was the focus of the first wave of mass participation in tourism during the nineteenth century, the second wave of mass tourism pushed its peripheries to foreign countries. Different stages of development can be recognised post-1950, with the latest periphery being space. New types of tourism emerged towards the end of the last century, including ecotourism and sustainable tourism. There is also a trend towards independent travel, booked on the Internet, rather than using tour operators and travel agents. This is representative of an increasing familiarisation with the experience of tourism in many societies.

Suggested reading

Bray, R. and Raitz, V. (2001) *Flight to the Sun: The Story of the Holiday Revolution*, Continuum, London.

Brendon, P. (1991) *Thomas Cook: 150 Years of Popular Tourism*, Secker and Warburg, London.

Davidson, J. and Spearritt, P. (2000) *Holiday Business: Tourism in Australia since 1870*, Melbourne University Press, Victoria.

Dumazedier, J. (1967) *Towards a Society of Leisure*, Free Press, New York.

Inglis, F. (2000) *The Delicious History of the Holiday*, Routledge, London.

Hall, C.M. and Page, S.J. (1999) *The Geography of Tourism and Recreation: Environment, Place and Space*, Routledge, London.

Towner, J. (1996) *An Historical Geography of Recreation and Tourism in the Western World: 1540–1940*, John Wiley & Sons, Chichester.

Suggested websites

World Tourism Organisation   www.world-tourism.org
World Travel and Tourism Council www.wttc.org

SOCIOLOGY AND TOURISM

2

This chapter will:

- consider how sociological theory can be applied to tourism;
- critically discuss the influence of industrialisation upon the motivation to travel;
- explore how tourism can be interpreted as a commodity;
- consider issues of social exclusion from participation in tourism.

Introduction

Chapter 1 focused on a historical and geographical account of tourism, highlighting the social and economic changes in place that help explain contemporary patterns of tourism. Having established that tourism has become an integral part of global society, the focus of this chapter is upon attempting to understand the significance of tourism in contemporary society. Based upon the account of tourism in Chapter 1, it is suggested that tourism may be viewed not only as a matter of individual preference, but also as a type of behaviour that is reflective of the societies that tourists come from, and that impacts upon the societies that they visit. In attempting to understand the role of tourism in society, one social science discipline that has a special usefulness in this task is sociology.

The origins of sociology

The term 'sociology' originates from the Latin *socius* meaning a companion and the Greek *logos* meaning 'study of', hence it literally means the study of the process of companionship (McLeish, 1993). Particularly, its origins reflect a desire to attempt to understand the social forces that structure society and influence behaviour. Commonly acknowledged as the founding father of sociology is Auguste Comte (1798–1857), born in France, who was the first person to attempt to move the understanding of society from a philosophical to a scientific basis.

To give a context as to why Comte was interested in trying to establish a scientific study of society, it is necessary to consider the social and economic conditions of early nineteenth-century France. It is not a coincidence that Comte lived at a time of tremendous social upheaval as a consequence of the French Revolution in 1789. The poignant impact of the beheading of Louis XIV has been likened to the killing of 'God', on the basis that the monarchy was established upon the principle of divine rule. Consequently, the French Revolution challenged the social structure of not just French, but the whole of European society. Additionally, the effects of the Industrial Revolution were also being felt in Europe at approximately the same time.

Comte was particularly concerned with how to establish a reliable knowledge of human behaviour that could provide the basis for improvements in social welfare (Giddens, 2001). Drawing upon inspiration from research in the natural sciences, Comte attempted to establish a 'science of society', which he originally termed 'social physics', and later renamed as 'sociology' in 1838 (Slattery, 1991). Comte's view was that all human behaviour was the result of observable forces external to the individual, a continuance of the philosophical approach of naturalism explained in the Introduction. He believed that there were laws that explained the patterns of the social world just as there were scientific laws that governed the natural world. He termed this theory 'positivism', to emphasise the positive outcomes that would result for society through the establishment of these laws. Given the context of social change and flux in France in the first half of the nineteenth century, it can therefore be argued that sociology emerged as a conservative ideology, which offered an image of society based upon the concept of order in contrast to the surrounding social chaos (Dann and Cohen, 1991).

Inherent to the positivist approach founded by Comte is that sociology should only be concerned with observable entities that can be objectively measured, and that such measurement is essential to explain behaviour. Consequently, positivism rejects the 'abstract' and 'subjective' aspects of human nature, such as emotions and feelings, favouring observation, categorisation and the measurement of 'facts'. Thus, adopting a positivist approach to understanding tourism as a fact of society would typically be based upon measuring different aspects of it, in an attempt to establish the social laws that govern it. The rationale for the ignoring of emotions, interpretations and feelings is that not only are they unable to be measured effectively, but also that they may distort any objective analysis. Although positivism was influential as a methodology to establish the laws of sociology up to the late 1950s, it has subsequently been challenged as a method for understanding society. The next section of the chapter considers the main theories of sociology which lay claim to explain how society works and develops.

Theories of sociology

Structuralism

In sociology, positivism is now incorporated within the broader heading of structuralism, which views society as existing prior to individuals (Abercrombie *et al.*, 2000), consequently an individual's beliefs and behaviour can be interpreted as being controlled by the rules of society (Sharpley, 1999). Structuralism, as the name suggests, analyses the structures of society, how they fit together, and in turn how they influence our behaviour.

Within this broad concept, two main schools of thought are evident: 'functionalism' and 'conflict theory'.

Functionalism

Within a functionalist perspective, society is viewed as a complex system whose various parts work together to produce stability and solidarity (Giddens, 2001). Social institutions such as the family and the Church are therefore analysed as part of a social system and are understood in terms of the contribution they make to the system as a whole. In the case of tourism, it would subsequently be analysed in terms of the role it plays in society and in the wider social system, rather than being treated as a separate activity that has no relationship to other parts of the social system.

Using the analogy of the human body, functionalists liken the workings of society to the working of an organism (Giddens, 2001; Haralambos and Holborn, 1990). Just as the efficient working of the various organs of the body in an interrelated fashion is necessary for its well-being, functionalists argue that the various parts or items of society work together in a similar way for the benefit of society as a whole. Subsequently, as Haralambos and Holborn (1990: 768) comment: 'an understanding of any part of society requires an analysis of its relationship to other parts, and most importantly, of its contribution to the maintenance of society'.

Functionalists also emphasise the importance of a moral consensus as part of maintaining order and stability in society, which they regard as its normal state, and they argue that consensus exists when most people in a society share the same values. A major figure of functionalism and positivism, and regarded as the key figure in establishing sociology as an academic discipline, is Emile Durkheim (1858–1917). Similar to Comte, Durkheim was French, and also believed that social life should be studied with the same objectivity with which scientists study the natural world. Durkheim argued that society has a reality of its own, and that individuals are controlled by 'social facts' that exist externally to the individual. Social facts are social forces, such as the state of the economy or the influence of religion, which compel people to follow certain patterns of behaviour.

Think point

Tourism is a form of behaviour in society. To what extent do you think 'social facts' or social forces exist in society that explain why people become tourists? Consider the economic, social and technological changes associated with the Industrial Revolution that were discussed in Chapter 1.

Conflict theory

The second major theory of structuralism, 'conflict theory', also emphasises the importance of structures within society but rejects functionalists' emphasis on consensus, instead highlighting the importance of divisions in society. As Giddens (2001: 17) states:

A common critique of functionalism is that it unduly stresses factors that lead to social cohesion, at the expense of those producing division and conflict. The focus on stability and order means that divisions or inequalities in society – based on factors such as class, race and gender – are minimised.

In contrast to functionalists, conflict theorists concentrate upon issues of power, inequality and struggle, interpreting society as being composed of different groups pursuing their own interests. Consequently, the pursuit of separate interests means that the potential for conflict is always present and that certain groups in a society will benefit more than others. According to Sharpley (1994), the values and norms of a society are founded upon the ability of a dominant group in society to impose their values and behaviour onto subordinate groups, in turn enabling them to maintain their dominate position. Subsequently, conflict theory seeks to examine tensions between dominant and disadvantaged groups within society, seeking to understand how relationships of control are established and perpetuated.

The development of conflict theory is especially associated with Karl Marx, who, although it can be argued that he was as much a political economist as a sociologist, is undoubtedly one of the key influences on modern sociology and society (Slattery, 1991). According to Giddens (2001), Marx interpreted differences of interest and conflict in society as a consequence of class differences and inherent power relationships. That is, in all societies there is a difference between those who hold authority and power and those who are largely excluded from it; between the rulers and the ruled.

From the perspective of conflict theory, the tourism system could be viewed as having a variety of power struggles taking place within it. These include concerns over the influence and power of transnational tourism corporations in lesser developed countries; the denial of resources by the powerful to the less powerful, e.g. villages being denied water for agriculture because it is needed to irrigate golf courses for tourists; the use of indigenous cultures against their wishes by central government to promote tourism, as is described in Box 6.2; and the power and gender dynamics of sex tourism.

Think point

Can you think of conflicts that can arise between different stakeholders within the tourism system? Why do these conflicts occur? How could these be explained within the context of conflict theory?

Phenomenology

The positivism that is an inherent part of structuralism, in both functionalism and conflict theory, has increasingly been attacked since the 1960s. A major criticism of structuralism is its treatment of humans as being 'governed' by invariable laws. In contrast to functionalism and conflict theory which are referred to as 'macro theories', because of their concentration upon offering an explanation of society as a whole (Haralambos and Holborn, 1990), phenomenology is an alternative sociological paradigm to knowing and understanding the world. Although phenomenology incorporates a diverse range of ideas, a basic agreement rests upon an emphasis of the study of the social world as being fundamentally different to that of the natural world. Subsequently, the focus of phenomenology is on how we give meaning to and interpret the world (Harvey *et al.*, 2000), rather than trying to discover the social laws and facts that explain and govern it.

The founding of phenomenology in a sociological context is associated with Edmund Husserl (1858–1938), who advocated that it is impossible to say anything very certain about the external world (Slattery, 1991), and was keen to demonstrate the falsity of the assumed separation of scientific knowledge from people's experiences and actions (Harvey *et al.*, 2000). Phenomenologists argue that society does not 'exist', but is created through routine, human interaction and shared assumptions (Slattery, ibid.). Emphasis is therefore placed on trying to understand how an individual interprets and creates their world through perceptions, motives, feelings, imagination and other mental processes. A further major difference to positivism is that emphasis is placed upon the ability of the individual to be able to influence and control their own world, rather than being controlled by external or causal effects.

Influential in this line of thinking was Alfred Schutz (1889–1959) who, in his theory of 'social action', viewed people as active agents who create and shape their society. Consequently, an assumption of social action theory is that the action is meaningful to the individual, therefore understanding the action requires an interpretation of the meanings that actors give to their activities. Thus, from a social action perspective, tourism is a meaningful behaviour and to understand it requires that we attempt to get into the mind of the tourist. Therefore reliance is placed upon an interpretation of the consciousness as Haralambos and Holborn (1990: 19) note: 'Since it is not possible to get inside the heads of actors, the discovery of meaning must be based on interpretation and intuition.' Instrumental in developing an interpretive approach of human action was Max Weber (1864–1920), who, in his concept of 'Verstehen', emphasises placing oneself in the position of other people to see what meaning they give to their actions.

Phenomenologists argue that the most sociologists can hope to do is to understand the meaning that individuals give to a particular phenomenon (Haralambos and Holborn, 1990). Emphasis is therefore placed not on searching for the explanations of a phenomenon, for example crime or tourism, but upon understanding how crime or tourism is defined. The end product of phenomenological research is thus an understanding of the meanings employed by members of their society in everyday life. Hence in the case of tourism, within a phenomenological perspective, emphasis would be placed upon attempting to understand the meaning and definition of tourism held by individuals, rather than trying to establish the rules of society that govern participation within it.

However, phenomenology has been criticised for its lack of scientific rigour and a subsequent subjectivism in its interpretation. Additionally, its research projects have been criticised for being small scale, concentrating on small group activity and interaction, and lacking the development of theory to analyse the whole of society (Slattery, 1991). Subsequently, neither of the major sociological paradigms of functionalism or phenomenology provide definitive theories of how to understand the world, or, for that matter, tourism. They illustrate that trying to understand and make sense of what is happening around us remains a contentious philosophical issue.

> **Think point**
>
> What are the main differences between a phenomenological approach to understanding tourism compared to a structuralist perspective?

The application of sociological theory to tourism

While the sociology of tourism is relatively underdeveloped compared to other areas of sociological enquiry (Cohen, 1984; Urry, 1995), and no single theological perspective can reasonably claim a monopoly in providing an understanding of tourism (Dann and Cohen, 1991), possible areas for a sociological enquiry of tourism are shown in Figure 2.1.

The diversity of the possible areas of enquiry, and the ability of other social science disciplines to make equally valid claims for investigation, led Sharpley (1994: 19) to comment: 'what is the most appropriate "ology" to use?' Alternatively, it could be asked what route should a sociological enquiry of tourism pursue? In the view of Dumazedier (1967: 124): 'vacation travel is linked with the degree of urbanisation of a country.' He continues to suggest that sociology should focus on the conditions of the leisure migration between home and vacation spots. Similarly, Sharpley (1994: 24) comments that: 'The greatest potential for tourism research within the holistic, macro approach is in adopting the basic assumption that tourism is a reflection of society.' Subsequently, the approach taken in this chapter is to apply sociological theory to the analysis of the key themes of urbanisation and industrial development, which were developed in Chapter 1 to explain the growth in contemporary tourism. As Hall (1994: 192) suggests: 'Tourism

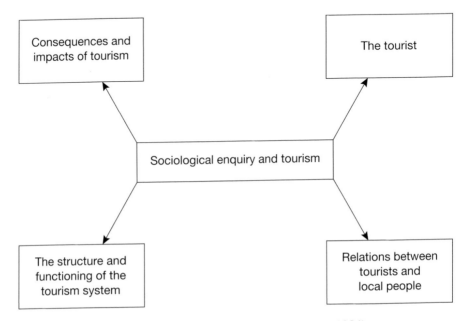

Figure 2.1 Areas of sociological enquiry in tourism (after Cohen, 1984).

grew out of the changing politics of space and time as the social and economic impacts of industrial revolution took place in the nineteenth century.' He continues: 'Tourism is a product of capitalism . . . one approach to understanding contemporary mass tourism is as an activity of society and cannot be understood with reference to it.' Consequently, the next section of this chapter considers sociological theories that have a direct application to understanding how society influences tourism.

The blasé and anomie

The first notable attempt to understand how urbanisation affected society was made by George Simmel (1858–1918), a strong advocate of social action theory, who observed the lives of the population of Berlin at the beginning of the twentieth century. Simmel's observations took place at a time of a rapidly growing Berlin, with high levels of immigration especially from Poland, which led to Berlin becoming the third most populated city in Europe after London and Paris. In his essay 'The Metropolis and Mental Life' (1903), he put forward ideas about the interaction between individual consciousness and the city (Lechte, 1994), viewing society as being created through a web of interactions between people.

In terms of how urbanisation affects the individual, he suggested that ties between people become more formalised, replacing more traditional affective ties (Lechte, 1994). A key aspect of this formalisation is money, which acts as the medium of exchange for a wide variety of goods. However, Simmel's major insight into how urbanisation affects the individual consciousness is to suggest it brings about a new psychological disposition (Lechte, ibid.). Faced with an over-stimulation of the senses from the complexity and diversity of life in urban areas, it is necessary to develop attitudes of reserve and insensitivity to feeling, which can be termed the 'blasé'. Without the development of the blasé, people would not be able to cope with the experiences resulting from high population densities.

Based upon his observations Simmel (1903: 318), cited in Bocock (1993: 16), states: 'the deepest problems of modern life derive from the claim of the individual to preserve the autonomy of his existence in the face of overwhelming social forces.' One route to a type of autonomy is through participation in tourism, which by taking place away from the routines of work and home may offer time that is free from the controls and pressures of daily life. One possible explanation of tourism could thus be as a reaction to the over-stimulation, manifesting itself in Boorstin's (1961) assertion of tourism as a form of mindless escapism, or in MacCannell's (1976) view of tourism as a search for 'authenticity' that is absent in everyday life.

Think point

Many people who live in cities develop attitudes of reserve and insensitivity as a way of coping with over-stimulation, which Simmel termed the blasé. How could the blasé and a search for autonomy or self-direction manifest itself in tourism?

In contrast to Simmel's interactionist perspective of urbanisation, Émile Durkheim (1858–1917) used a structuralist perspective to understand the effects of urbanisation upon the individual. A central thesis of Durkheim's was that a society was held together by the sharing of a set of social guidelines or norms, based upon a moral consensus or collective consciousness (Slattery, 1991). These norms set the boundaries of acceptable behaviour within a society or culture. In more traditional pre-industrial cultures, he believed that members of society were bound together by common experiences and lifestyles, with shared beliefs that evolved from having similar occupations, which he termed 'mechanical solidarity'. There existed what Durkheim referred to as a 'conscience collective', that is a collective morality or set of values which guide and control individual behaviour (Slattery, ibid.). The strength of these shared beliefs is repressive to any individual action which threatens or challenges the community's traditions and existing patterns of life.

In Durkheim's view the processes of industrialisation and urbanisation threatened this solidarity, as labour was divided into more specialist types and tasks, and people relocated from rural to urban areas. Consequently, the commonality of lifestyle that was a characteristic of pre-industrial communities began to be replaced by differentiation of employment, and by individualisation. The face-to-face relationships that characterised mechanical solidarity, and other forms of informal social controls including cultural customs that held society together, were replaced by more formalised ones through the state and the law. Thus, a new basis for the organisation of society is required that is capable of combining individual freedoms with social order, which Durkheim termed 'organic solidarity'. Subsequently, mechanical solidarity has been replaced by 'organic solidarity', where social cohesiveness is dependent upon an individual's economic interdependence and a more formalised social order provided by the state (Giddens, 2001).

In Durkheim's view, the chance of a breakdown of a social consensus and of social controls over the individual was more likely in societies held together by organic solidarity (Slattery, 1991). He observed that the norms and solidarity of society were most at risk during periods of social upheaval or transition, such as during political revolutions and industrialisation, which would ultimately lead to feelings of aimlessness or despair. In his concept of *anomie*, a Greek word meaning 'lawlessness', Durkheim attempted to explain individual behaviour in terms of the wider social structure. According to Durkheim, anomie exists where there is an absence of clear standards and rules to guide behaviour in social life (McLeish, 1993), and in such a situation he asserted that individuals would feel threatened, anxious and disorientated. He believed that anomie was pervasive in modern societies and was a social factor that contributed to suicide. Durkheim observed that rates of suicide increased during times of economic depression and economic boom, as he argued both were times of rapid social change, disrupting the stability of lives of individuals. The theory of anomie was applied to tourism by Dann (1977) in one of the first major empirical studies of tourism, as is described in Box 2.1.

Work and alienation

A major interest of sociologists is the influence of work and employment upon society. Patterns of contemporary employment have been defined by the Industrial Revolution, becoming more formalised, time conscious and spatially separated from home than had

Box 2.1 Anomie and tourism

The linking of the theory of anomie to tourism was made by Dann (1977). In one of the first major empirical studies of tourists, Dann emphasised the factors of escapism and searching for social status as principal reasons for people becoming tourists. He interviewed 422 tourists to Barbados, to discover their reasons for choosing their holiday. Based upon a sociological interpretation of tourism motivation, Dann adopted the socio-geographical terminology of 'push' and 'pull' factors, to explain factors which encourage or motivate us to leave our home for another destination (push), and attributes of a destination that attract us towards it (pull). According to Dann, pull factors can only have validity after the push factors have already made up the mind of the individual to travel. He suggests that the key motives for travel, or push factors, lay in the twin concepts of 'anomie' and 'ego-enhancement'. The primary motivations resulting from the state of anomie are associated with emphasis being placed on social interaction with family or friends, and a search for meaning.

Dann's other key concept of ego-enhancement is based upon the individual's need for social recognition. If people are denied status at home, or, alternatively, individuals perceive themselves as having a low status, travel offers the opportunity to act out a new role even if this is for a transitory period. Ego-enhancement may also be achieved not only within the destination, but also upon the return home by having been to a fashionable destination or pursued the right type of holiday activity. The propensity for achieving ego-enhancement through tourism is likely to correlate with an increase in the wealth differential between the tourist and the destinations they are visiting. For example, travelling to destinations in lesser developed countries provides western tourists with the opportunities to purchase a range of services and goods that they could not afford at home. However, this wealth disparity also exposes tourism to accusations of 'servility', and may cause resentment in the local population.

Think point

How important is social interaction with family or friends as a reason for taking a vacation or holiday? How does tourism offer a pattern of interaction that is different to the one at home? Has tourism ever helped enhance your ego or self-confidence? Is so, how?

---

been the case in pre-industrial societies, which in turn has had individual and social consequences.

For Marx, work represented the most important of all human activities, being the medium through which human beings had the opportunity to express their individuality and creativity. Work may subsequently either provide a means of fulfilling our individual potential, or alternatively, distort and pervert relationships with ourselves and with each other (Haralambos and Holborn, 1990).

Marx argued that the modern worker of the nineteenth century had, through the processes of capitalism and industrialisation, lost much of their control of work that they possessed in more traditional societies. Notably, they now produced goods for a distant market rather than for themselves or for specific customers, and the division of labour associated with factory production meant that workers no longer produced a complete product, unlike traditional craftsmen. In Marx's thesis, the combination of workers' powerlessness, their loss of autonomy, and the loss of ownership and control of the final product leads to a state of 'alienation', which is both unsatisfying and unrewarding. Marx stated that as workers in capitalist societies become alienated from their work, given that work is the primary human activity, they become alienated from themselves and, ultimately, each other. Consequently, self-interest becomes more important than concern for a wider social group.

Although all previous economic systems that predated capitalism were based upon class structures and exploitative relationships, Marx argued that capitalism increased alienation not only because exploitation is at its highest, but also because the market mechanism encourages the treatment of workers as commodities for production, to be used and discarded as required (Slattery, 1991). Thus, in Marx's view, workers in a capitalist industrialised system have less autonomy and fulfilment and are subject to market forces in terms of job security. Although Marx was writing in the nineteenth century, much contemporary employment displays similar characteristics of a lack of autonomy and job insecurity.

Given in Marx's analysis that work is central to our lives, the fulfilment, or, conversely, the lack of fulfilment experienced through it, will affect other aspects of our life. This theme was developed by Krippendorf (1986), who remarked that tourism is not a separate world governed by its own laws, viewing it as a result and also as a component of the industrial system. Krippendorf views tourism very much as a phenomenon of opposition to that of everyday life in industrial and post-industrial societies, commenting (1986: 522): 'One seizes every opportunity to free oneself: to escape from the boredom of everyday life as often as possible. . . . Above all, one does not want to stay at home but to get away at any price.' He suggests that people travel because they no longer feel at ease where they are, neither where they work nor where they live.

Directly linking Marx's theory of alienation to leisure, the French sociologist André Gorz argued that alienation at work leads the worker to seek self-fulfilment in leisure (Haralambos and Holborn, 1990). However, critically in Gorz's view, leisure is also controlled by capitalism rather than having the more spontaneous community base of pre-industrial cultures. Leisure subsequently adopts the identity of an entertainment industry, ready for consumption, which in Gorz's view is a poor substitute for self-directed and controlled leisure. Therefore, leisure simply provides 'a means of escape and oblivion', a means of living with the problem rather than providing a solution to it (Haralambos and Holborn, 1990: 315). Similarly, the early sociology commentary of tourism, such as that from Boorstin (1961) and Dumazedier (1967) emphasises tourism as an expression of escapism and freedom. Yet the influence of this 'escape' through tourism on the home society is uncertain as Dumazedier (1967: 126) comments: 'It would also be interesting to know if vacation travel helps one adapt to city life, or, on the contrary makes such adaptation so difficult as to lead to chronic dissatisfaction or outright departure.'

Think point

To what extent do you think tourism offers a form of escapism? Does it make you feel more or less satisfied with your life when you return home?

The theme of the use of leisure and by implication tourism, as a means to support the model of the advanced industrial system, was an important line of thought in the Frankfurt School of sociology and notably Herbert Marcuse. The Frankfurt School situated in the Institute of Social Research at Frankfurt University was the first Marxist-orientated research centre to be affiliated to a major university. It aimed to provide a more humanistic path to socialism and a 'free' society. The Frankfurt theorists wanted to show that reality wasn't 'real' but that it was an ideological distortion to conceal and legitimise the power of the ruling class.

This association of neo-Marxists which composed the Frankfurt School was instrumental in developing a line of sociological enquiry known as 'critical theory', which analysed all forms of domination. They were highly critical of positivism because they viewed it as presenting the what 'is' in society as opposed to the what 'should be', thereby legitimising the existing social order and obstructing change. In contrast they proposed 'dialectical theory', i.e. taking a critical approach to all knowledge, as the basis for promoting pure reason and for liberating human thought. By never accepting any argument, fact or theory as totally proven, dialectical theory hoped to provide a basis for both criticising and changing the world. Emphasis was placed upon the freedom of the individual against the controlling forces of bureaucracy, technology, the media and the state.

It is their view on leisure that makes the Frankfurt School of sociology particularly relevant to tourism. In his book *One Dimensional Man* published in 1964, Marcuse advocated that leisure is based on and directed by 'false needs', largely imposed by a mass-media controlled by the establishment (Haralambos and Holborn, 1990). In using the term 'false needs', Marcuse meant that needs were false if they did not achieve a final state of self-fulfilment and real satisfaction. The result or end product of the fulfilment of false needs is a 'euphoria in unhappiness', a feeling of elation built onto a foundation of misery.

In Marcuse's view, while western society has become wealthier and created the possibility for a wider participation in leisure and tourism, 'chains of iron' have been replaced by 'chains of gold'. That is, the ruling classes and ruling elites having strengthened their hold over the workforce by making its exploitation more bearable. The fulfilment of false needs thus serves to divert attention from the real source of alienation; the nature of work. Thus it could be argued that tourism, for example the parade at Disneyland shown in Figure 2.2, far from providing a sense of real fulfilment does little more than satisfy a false set of needs to help make our lives more bearable.

The influence of work upon leisure was also viewed as being critical by Parker (1983), defining leisure as a residual category of time left after work, including not only paid employment but also other obligations such as housework and childcare. In his view, leisure activities are conditioned by work, stressing that the degree of involvement

Figure 2.2 The parade at Disneyland, California.

Think point

Figure 2.2 shows a parade at Disneyland in California. To what extent do you agree with the proposition that tourism can be viewed as a type of 'chains of gold'? Does the tourism industry provide little more than 'pseudo-events', offering inauthenticity and the fulfilment of 'false needs'?

---

people have. and intrinsic satisfaction gained in employment, directly influences their choice of leisure activities with three main patterns being recognisable. In the 'extension' pattern, the division between work and leisure lacks definition, and activities in both areas are similar. People who belong to this category are likely to have high levels of autonomy and involvement at work, and subsequently gain intrinsic satisfaction through their employment. Conversely, in the 'opposition' pattern, people dislike work so intensely that the quality of their leisure is defined by how unlike work it is. Typical characteristics of such employment would typically include limited self-autonomy and self-determination. The final type of relationship is that of 'neutrality'. In this case the time and energy spent in either sphere has no direct influence upon the other. However, there exists little empirical research on how employment influences tourism behaviour, either to support or question this model.

Think point

To what extent do you think work influences the types of holidays people take? Think of family and friends, their employment and the types of holidays and vacations they take. Do any patterns emerge similar to Parker's (1983) classification?

Commodification and semiotics

For the mass participation in tourism to have taken place in the latter part of the nine-teenth century, a large capital investment in a tourism infrastructure was required, including railways, steamships, hotels and seaside amusements. Combined with rises in levels of disposable income and the segregation of leisure as time spent away from work, it was inevitable that leisure and tourism would become commoditised, to be exchanged in the market. It was because of this process that millions of people were able to partic-ipate in tourism who had been unable to do so before. It can subsequently be argued that capitalism was essential for making tourism available for the masses rather than the select few in society.

The fact that tourism is bought and sold in the market means it is possible to think of tourism as a commodity for consumption. For Watson and Kopachevsky (1996), thinking of tourism in the context of contemporary consumer culture is the best way to understand it; as an extension of the commodification of life. Tourism can thus be interpreted as a form of consumerism, having similarities to buying a car or clothes. Similar to the buy-ing of cars and clothes, the reasons for purchasing tourism may be far more complex than purely a functional need, involving the construction and defining of one's own identity.

From our own experiences, we know that commodities have a value that passes beyond their power to satisfy human needs or, as Marx termed it, their 'user-value'. Most products carry with them signs, a 'sign-value' that transmits messages about it that pass beyond its use as a product. For example, while a Porsche is used as a car, it also suggests success, power and wealth. The study of these signs is known as 'semi-ology', and given that tourism can be viewed as a commodity, it can aid our under-standing of tourism as a type of social behaviour. For example, a holiday to Mauritius from Europe, for many people, would be likely to be interpreted more complexly than purely a holiday, suggesting attributes of wealth, high status and success.

Thus by participating in certain kinds of tourism or by travelling to particular places we construct an identity and also transmit messages about ourselves in society, as Mowforth and Munt (1998: 132) comment: 'It is symbolic in the way in which travel and tourism embody certain attributes: personal qualities in the individual, such as strength of character, adaptability, resourcefulness, sensitivity or even "worldliness".' Consequently, tourism can be used as a means of differentiating oneself from other types of people.

The use of tourism as a form of social differentiation can be traced at least to the onset of the Industrial Revolution, with observations of the potential symbolism of tourism being made by Veblen (1899) in late nineteenth-century America. Witnessing an emerging middle class in America becoming wealthy through trade and manufacturing, he noticed how they were using leisure and tourism to differentiate themselves from other citizens. The financial costs of international tourism limited participation to a privileged elite. Consequently, by sending one's wife or daughter on holiday from America to Europe, a message of wealth was conveyed to wider society. This 'conspicuous consumption' as Veblen (1899) termed it, in this case a conspicuous exhibition of exemption from the constraints of employment, illustrates how the semiotics of tourism can be used as a means of social differentiation.

The need for social differentiation in society is also a theme of the works of the French sociologist Pierre Bourdieu. Two key concepts of Bourdieu's, those of *habitus* and 'cultural capital', relate to a desire for social differentiation. Lechte (1994: 47) explains *habitus* as being: 'a kind of grammar of actions which serves to differentiate one class (e.g. the dominant) from another (e.g. the dominated) in the social field'. Cultural capital can be understood as the acquirement of a significant amount of knowledge, understanding and 'taste', which is not formally learned but unconsciously acquired through the family environment. The combination of *habitus* and cultural capital thus provides the means for social differentiation within a population.

Encompassed within this 'grammar of actions' of *habitus* is the symbolism associated with the consumption of tourism. As Mowforth and Munt (1998) point out, as consumers have become increasingly sensitive to the symbols they are consuming, the consumption of holidays has assumed a significant role in social differentiation. Social differentiation through tourism may be characterised by at least three variables: (i) those who can and cannot participate in it; (ii) the destinations where people choose to go; and (iii) the type of tourism activities they pursue while on holiday.

As has been previously stated, a basic requirement to participate in tourism is the ability to have the disposable income to spend on it. Yet, even if everyone had equal financial means to participate in tourism, we would be likely to choose different destinations and types of holidays, our choices carrying symbolic meanings and messages about ourselves to others. Besides the type of holiday and destination, social differentiation also exists in the types of transport used for tourism. Urry (1990) points out that even when the railways became more democratised in the nineteenth century, distinctions between the class of traveller remained, with the use of first, second and third classes. He argues that similarly this theme of status extends to other modes of travel, e.g. air travel can be segregated by first/business/economy classes and by scheduled/packaged flights. A summary of how tourism can be used for social differentiation is shown in Figure 2.3.

Think point

The following is a list of destinations and types of holidays: Miami Beach in Florida; Cannes on the Cote d'Azur; Bondi Beach in Sydney; package holidays; space tourism; and ecotourism. What do these places and types of holiday signify about the people who visit and participate in them?

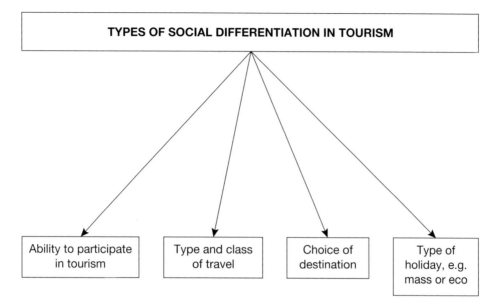

Figure 2.3 Types of social differentiation in tourism.

Social exclusion and marginalisation

As tourism has become an increasingly significant and mainstream component of many economically advanced societies it has also become more of an expected experience rather than a luxury one. An inability to go on holiday may therefore result in a lack of opportunity to participate in the mainstream lifestyle of a community. Reflecting the growing importance of tourism as a feature of society, Urry (1995) links access and participation in tourism with citizenship. He comments (1995: 30): 'Being able to go on holiday, to be obviously not at work, is presumed to be a characteristic of modern citizenship which has become embodied into people's thinking about health and well-being.' Consequently, exclusion from participation in tourism may be viewed as a denial of status and citizenship. Subsequently, the sociology of exclusion from participation in tourism, whether by class, sexuality or race, is becoming an emerging theme of the sociology of tourism.

As was explained in Chapter 1, tourism is usually defined from a western perspective, with a limited understanding of the meaning and significance of tourism in other cultures. One exception to this statement is a study of the Afro-Caribbean community in Britain described in Box 2.2.

Similarly, initial research conducted with the Pakistani community in the UK also suggests that the conception of tourism by this ethnic minority is very different from the mainstream western one. There exists no direct translation of the word 'tourism' and travel is undertaken either to visit friends and relatives in Pakistan, often where the spiritual centre of existence is felt to be, or a religious pilgrimage (Ali, 2004). The influence of religion on travel within this community is strong, with every Muslim

Box 2.2 Conceptions of tourism in the Afro-Caribbean community in Britain

In research with the Afro-Caribbean community of Manchester in England, Stephenson (2002) found the meaning and interpretation of tourism to be substantially different to that of the white population. Far from travel being something undertaken for pleasure, it was viewed as having associations with the exploitation of the transatlantic slave system, and more recently the necessity to relocate from the Caribbean to the UK as a form of economic migration. The concept of the 'holiday' was regarded as a western white activity, while for the Afro-Caribbean community, travel was almost exclusively restricted to visiting friends and relatives in Jamaica. This was particularly the case for first and second generation Afro-Caribbeans, whose tourism aspirations focused overwhelmingly on the 'ancestral homelands', despite having lived in the UK for the past five decades.

being expected to undertake Hadj at least once in their life, and consequently visit Mecca in Saudi Arabia.

However, the socio-cultural boundaries that differentiate ethnic groups are also in the process of shifting and to a certain extent disintegrating, as individuals become interested in others' lives. In societies where there is a high ethnic mix, and that are also subject to cosmopolitan and global influences such as through media and fashion, it could be expected that the desire to travel will expand beyond the more traditional horizons associated with culture, history and family. Research conducted with young and well-educated Pakistanis in Britain by Klemm (2002) supports this hypothesis. She found their holiday preferences were not substantially different from the British population as a whole, for example the most important motivations for going on holiday were 'break from work' (65 per cent) and 'seeing a new place' (50 per cent).

Klemm (ibid.) also found that the marketing and promotional techniques used by the mainstream tourism industry were perceived as being negative by this group. It was felt that holidays were for 'white' people only, being defined by characteristics of a lack of ethnic minority representation in tour operators' publicity; a lack of consideration of food that would be required for Muslims' dietary requirements; and an emphasis placed upon beach holidays which had little appeal to the Pakistani community.

A theoretical framework that can be applied to explain a lack of ethnic minority participation in tourism was developed by Washburne (1978). The 'marginality hypothesis' suggests that the travel preferences of ethnic minorities are constrained by their marginal position in society and low incomes. Subsequently, the discriminations and restricted opportunities that exist in many areas of life for ethnic minorities, spill over into tourism. In contrast the 'ethnicity hypothesis' suggests that tourism preferences are largely determined by family background and cultural identity.

Barriers to travel

Lack of income

For any individual or group an obvious barrier to participating in tourism is a lack of disposable income to be able to purchase it. According to the European Commission Directorate General (1998) the most common reason given by European citizens for non-participation in tourism was financial, ranging from a high of 66 per cent in Portugal to a low of 29 per cent in Luxembourg.

While there would seem to be a strong correlation between Gross Domestic Product (GDP) per capita and the proportion of people taking holidays, strong differences within countries may exist based upon structural features, especially social class. For example, in the UK, as Shaw and Williams (2002: 61) comment:

> those from social class groups D and E (including unskilled workers) account for 33 per cent of the population, they make up 49 per cent of the adults not taking holidays. Conversely, those in professional and managerial occupations (groups A and B) comprise 17 per cent of the population, but only 9 per cent of those not taking holidays.

Similarly, Seaton (1992) comments that higher socio-economic groups are much more likely to go on holiday than the lower socio-economic groups, with differentiation also being made between the destinations visited and the accommodation used. Based upon the population in the UK, upper-class social groups were more likely to visit France, Italy, US, Switzerland and the West Indies and stay in hotels and rented accommodation, whereas the lower social groups were more likely to visit Majorca, Minorca and Ibiza, staying in guesthouses, holiday centres and camping sites.

While disposable income may act as an inhibitor to participation in tourism, cultural inhibitors may also constrain the desire to travel. For example, Smith and Hughes (1999) found that economically disadvantaged families in the north of England felt uncomfortable and conspicuous when going on holiday. The reasons for this included an unfamiliarity with the expected 'rituals' and 'customs' of holidays, and also that their clothes and travel bags made them feel conspicuous.

Racism

Besides a lack of income, prejudices that marginalise minority groups in society, such as racism and homophobia, also spill over into tourism and inhibit travel, as is described in Box 2.3.

Travel to certain places may also be inhibited when they are perceived to have an identity that is hostile or discriminatory. For example, the countryside in England is presented very much as a place of 'white Englishness', established by symbolic images in the media, art and literature, which produces 'racialised boundaries' (Stephenson, 2004). Thus it is presented as a place of 'white safety', differentiated from urban areas and associated social problems. Perhaps this is most aptly symbolised by the then British

---

Box 2.3 'British Muslims fly into a hostile climate'

This was the title of a newspaper report following the attack on the Twin Towers in New York on 11 September 2001 and a reported rise in 'Islamophobia'. The basis of the report was how British Muslims and members of other Asian ethnic minority groups felt themselves to be increasingly on the receiving end of hostile treatment at the hands of immigration officials and aviation authorities. Innocuous reasons such as language and dress have become reasons for suspicion. Wazir (2001) comments that men of Arab origin were asked to disembark from a charter flight from Stockholm to the Canary Islands because other passengers were concerned about their presence. Similarly, authorities in Beijing told major Chinese airlines to stop ticket sales to nationals from 20 countries, especially those in the Middle East.

Source: Wazir (2001)

Prime Minister John Major's comment in 1993, that there will always be an England with 'postmistresses riding bicycles across village greens in the mist' (Neal, 2002: 444). The issue of a lack of ethnic minority representation in visits to the British countryside, and the consequential denial to individuals of the health and enjoyment benefits including those of a more natural landscape, has not gone unnoticed by the government as is described in Box 2.4.

A powerful example of how racism can act as an inhibitor to travel is given in Box 2.5 based upon the experiences of foreign-exchange students.

Think point

To what extent do you think tourism has become a part of citizenship? What kind of factors can result in exclusion from tourism? Have you ever experienced or witnessed racism as a tourist?

Homophobia

A further group for whom participation in tourism may be restricted by prejudice is homosexual or gay people. How tourism is used by the gay community and their experiences as tourists is becoming a growing field of tourism research. While there exists a growing interest from the tourism industry in attracting the gay and lesbian markets as they are held to have high levels of disposable income, often referred to as the 'pink pound', their choice of destinations is limited by cultural and social prejudices towards homosexuals.

The importance of holidays for some gay men is that it offers the opportunity for them to be themselves, freed from the constraints of their home societies in which homosexual activity may be marginalised. This point may be particularly important for gays who may not feel free to frequent 'gay space' at home because of their fear of discovery, but feel more comfortable to do so in the anonymity of the tourist destination. However,

Box 2.4 Lack of ethnic minority representation in the countryside

Concerned by the lack of a black ethnic-minority presence in the countryside in Britain, the Department for Environment, Food and Rural Affairs has asked the Countryside Agency how to boost the numbers of ethnic minorities visiting rural Britain. According to Benjamin (2004) many non-white people choose not to visit because they feel excluded, with no sense of ownership, and fear of abuse. For people from ethnic minorities born in the country, a lack of familiarity with countryside may also be a deterrent, having never been taken there by their parents. This is despite many people from cultures including the Jamaican, Indian and Pakistani having strong rural roots but whom have become stereotyped as 'inner-city' residents.

The extent of a fearful preconception of the countryside is emphasised by the reported reaction of one Asian ethnic-minority group to the countryside: 'Before visiting the countryside many Handsworth residents believed that they would be shot by the farmer' (Benjamin, 2004: 2). However, in this case a community group has actively worked with the community to organise trips to the countryside especially for children, which Benjamin (2004: 2) comments has: 'infected their parents with enthusiasm for rural life' and has also led to them demanding more green spaces in their own home towns. The issue of access to the countryside for ethnic minority groups is not just restricted to England, with similar issues being reported in the French countryside. It is likely to occur wherever there is a minority group and perceived or actual discrimination such as racism exists. Important in tackling this issue is the need to get ethnic minority groups actually involved in the participatory planning and development of the countryside, for example on the management boards of national parks.

Source: Benjamin (2004)

Box 2.5 'Unwelcome guests'

How racism can act as an inhibitor to travel is graphically described in the following passage about the experiences of a 14-year-old girl from a black ethnic minority on a student-exchange trip to Germany. Describing her experience of life with the family she was staying with she comments: 'The mother was making Tarzan shouts and monkey noises at me, and rolling and showing the whites of her eyes. . . . I felt like killing myself, I would have done anything to get away from there.' Similarly, the experiences of a French student on an exchange trip to Britain demonstrate how prejudice can influence the tourism experience. The father of the host family who was an elected representative of a mainstream political party, upon meeting the girl and seeing she was black, announced the arrangement as being 'no longer convenient'.

Source: Chaudhuri (1999)

Hughes (2002) comments that the sexual motivation of gay men as a reason for travel is stereotypical and overemphasised, with gay men having similar reasons for going on holiday to the rest of the population.

Not all destinations are keen on promoting themselves as gay destinations, a major reason being that this may deter the family market. For example, although Amsterdam is held to be the gay capital of Europe, a tourist-board campaign aimed at the US gay market in the early 1990s was not repeated, owing to the reaction of the tourist trade in the city. They felt that the campaign would create an 'unfavourable' image of the city and lead to it being marginalised (Hughes, 1998). Similarly, Want (2002: 194) citing a BBC report from 1998 on the docking of a cruise ship in the Cayman Islands comments:

> The government of the British territory of the Cayman Islands in the Caribbean has refused permission for a cruise liner carrying hundreds of homosexual holi-daymakers to dock there. The authorities claimed that there was no guarantee that, as they put it, the group would uphold appropriate standards of behaviour. . . . In a letter explaining the decision of the authorities, the tourism minister said landing rights were being denied because of what he called careful research and prior experience.

As Want (ibid.) observes, a country's legislation can be an indicator of the local climate towards gay and lesbian tourists, with, at the extreme limit, homophobia manifesting itself in the death penalty being legitimised by the state for homosexual activity. This would apply to some states in the Middle East, and in the major tourist area of the Caribbean many states have until recently retained 100-year-old laws, inherited from the British, that made homosexuality illegal. Notably, Jamaica has gained itself a reputation for having a high level of homophobia, which is reflected in the policy of the Caribbean resort chain Sandals to refuse all bookings from same-sex couples, a policy which has now been revised. Such attitudes are likely to have economic consequences, deterring heterosexuals from more liberal and progressive countries, besides alienating gay people. Conversely, the constitution of South Africa ensures that all tourism properties accept bookings from same-sex couples (Levitt, 2004).

As Hughes (2002) suggests, the choice of destination for the marginalised groups in society may be a much more constrained process than it is for heterosexual tourists, with a variety of inhibitors coming into action. Consequently, a priority may be risk-avoidance or minimisation, including lessening the chance of physical or verbal attack because of one's skin colour or sexuality. This could consequently lead to the establishment of a kind of 'tour' or an itinerary of 'safe' destinations to visit. In the case of the gay community this includes South Beach Miami, the Greek island of Myknos, the Spanish beach towns of Sitges, Ibiza Town and Gran Canaria, the cities of New York, San Francisco, Amsterdam, and the English seaside towns of Brighton and Blackpool (Hughes, ibid.). However, as Fearis (2004) observes, efforts are being made by specialist travel agents to extend the boundaries of 'pink tourism' to include Texan rodeos, Caribbean cruises, ski trips and city breaks.

The presence of inhibitors and barriers to participation in tourism has implications for the policy of the tourism industry and other agencies who may desire to encourage a

wider participation in tourism. Klemm and Kelsey (2002) point out that the tourism industry is predominantly orientated towards a white customer, while Makuni (2001) emphasises that many brochures and television advertisements selling holidays are orientated
towards the white nuclear family, ignoring other market segments. Subsequently, efforts
are needed for the tourism industry to accommodate minority-group interests. There is
also a strong economic rationale for the tourism industry to diversify its market, as Makuni
(2001: 78) points out: 'global trends have encouraged shifts in power resulting in the
emerging of groups which had previously been disregarded by mainstream society . . .
women, homosexuals, senior minorities . . . and ethnic minorities.' An initiative being
taken by one mainstream tour operator to diversify its market is described in Box 2.6.

Box 2.6 Thomas Cook and gay tourism

A reflection of how social changes are influencing the tourism industry is the decision by Thomas Cook to target the gay and lesbian market. They estimate the size
of the market to be two million in Britain alone and intend to develop packages not
just to tradional gay destinations such as Ibiza, Mykonos and San Francisco but also
to mainstream holiday destinations. Thomas Cook believes this is important as they
are aware that gay travellers want to be able to openly buy holidays and travel to
the same destinations as other tourists. Essential in achieving this, is for travel agents
and tour operators to make it clear that gay people are welcome and not stigmatised.
Subsequently, Thomas Cook is vetting owners of hotels, villas, ski chalets and cruise
ships to ensure that they offer gays a warm and non-discriminatory welcome.
However, as Levitt (2004: 1) comments: 'It takes more than a rainbow flag and a
Kylie Minogue soundtrack to make a location and a tour operator gay-friendly.'

After: Summerskill (2001)

Summary

- The origins of the sociology were associated with an attempt to move the
  understanding of society from a philosophical to a scientific base. From a
  'functionalist' perspective tourism can be viewed as a part of society that works
  with other parts to produce stability and solidarity. From the perspective of
  'conflict theory', the tourism system can be viewed as being part of and
  symptomatic of the wider power-struggles taking place in society. Tourism may
  also be viewed as a form of 'social action' in which individuals construct the
  meaning of tourism for themselves.
- Theories of 'anomie' and 'alienation' can be used to explain how industrialisation
  has created the economic and social conditions that encourage people to
  become tourists, as the 'mechanical solidarity' of society was replaced by
  'organic solidarity'. According to the Frankfurt School, leisure and, by

implication, tourism can be viewed as being based upon 'false needs', replacing 'chains of iron' with 'chains of gold'.

- For the mass participation in tourism to have taken place in the latter part of the nineteenth century, a large capital investment in a tourism infrastructure was required, including railways, steamships, hotels and seaside amusements. Combined with rises in levels of disposable income and the segregation of leisure as time spent away from work, it was inevitable that leisure and tourism would become commoditised. Tourism can subsequently be interpreted as an extension of the commodification of life and a part of contemporary consumer culture. It may subsequently be used to construct one's own identity as a form of 'conspicuous consumption' and form part of one's *habitus*, permitting social differentiation. Social differentiation through tourism may be characterised by at least three variables: (i) those who can and cannot participate; (ii) the destinations to which people choose to go; and (iii) the type of tourism activities they pursue while on holiday.
- Tourism may be viewed as a characteristic of modern citizenship associated with people's well-being. However, many people have this right to citizenship denied or restricted, through social exclusion from participation in tourism. Barriers to participation include poverty or a lack of money to spend on tourism; and wider social prejudices that exist in society, especially racism and homophobia.

Suggested reading

Giddens, A. (2001) *Sociology*, 4th edn, Polity, Cambridge.
Sharpley, R. (1999) *Tourism, Tourists and Society*, 2nd edn, Elm Publications, Huntingdon.
Slattery, M. (1992) *Key Ideas in Sociology*, Nelson, London.
Urry, J. (1990) *The Tourist Gaze: Leisure and Travel in Contemporary Societies*, Sage, London.

PSYCHOLOGY AND TOURISM

3

This chapter will:

- consider how psychological theory can be applied to tourism;
- critically discuss tourist motivation;
- evaluate personality as a predictor of tourist behaviour;
- describe how a tourist may experience the surrounding environment.

Introduction

While sociology can greatly aid our understanding of the forces in society that shape tourism and its significance, psychology can also enhance our understanding of tourism as a form of individual behaviour. Commenting on the meaning of psychology Davidoff (1987: 6) states: 'The word "psychology" is derived from the Greek word meaning "study of the mind or soul".' while Malim and Birch (1998: 3) comment that: 'Perhaps the most widely accepted definition of psychology is that of the scientific study of behaviour and experience.' Thus it can be said that psychology is a social science that focuses on behaviour and mental processes, attempting to provide a comprehension and understanding of human (and animal) behaviour. Yet, beneath this aim lies a myriad of complexity of terms and theories to help psychologists achieve this ultimate goal.

The interest in understanding human behaviour has a long history before the emergence of psychology with written records of speculation about human behaviour dating to the ancient Greek philosopher Aristotle (Davidoff, 1987). According to Davidoff (ibid.), the early origins of contemporary psychology can be traced to the work of the philosopher and physicist Gustav Fletcher (1801–1887), who researched into how scientific methods could be applied to the understanding of mental processes. Malim and Birch (1998) comment that psychology as a scientific discipline has a short history of just over 100 years; previous to this psychology had generally been regarded as a subset of philosophy. In the intervening period a diverse range of schools of thought of psychology attempting to explain human behaviour have developed, which are shown in Figure 3.1.

The variety of schools of thought of psychology shown in Figure 3.1 indicates the diversity of approaches to the subject. However, as Pearce and Stringer (1991: 137) comment:

Despite this diversity of detail in contemporary psychology, there are some broad goals and methods that unite the discipline. Most psychologists pay attention to the behaviour and experience of individuals and seek to describe and, if possible, explain any observed patterns in this behaviour and experience.

Subsequently, it would seem that psychology has a direct relevance to explaining tourism as a type of behaviour.

In terms of a historical application of psychology to tourism, Pearce and Stringer (1991) comment that some of the first scholars to call themselves psychologists were interested in visitor behaviour. Sir Francis Galton, a pre-eminent British psychologist in the early twentieth century, studied visitors to museums in London and established an observational laboratory in one of them to record human behaviour and reactions. In applying psychology to the study of contemporary tourism, Pearce and Stringer (1991) and Ross (1994, 1998) emphasise that there are core areas of psychology that have a relevance to tourism. These include motivation, personality, attitudes and environment, which are subsequently discussed in turn in the following sections of this chapter.

Motivation

The relevance of individual motivation to understanding tourism is poignantly summarised by Parrinello (1993: 233): 'The importance of motivation in tourism is quite

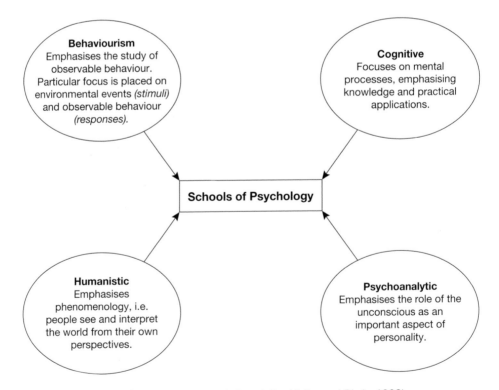

Figure 3.1 Schools of contemporary psychology (after Malim and Birch, 1998).

obvious. It acts as a trigger that sets off all the events involved in travel.' Axiomatically, without the desire and motivation to travel there would be no tourism system or tourism industry. However, agreement upon a psychological theory to understand motivation has not been reached (Atkinson *et al.*, 1983; Davidoff, 1994; and Gross, 1992) and neither has it in tourism research, as Pearce (1993: 114) comments: 'research into why individuals travel has been hampered by the lack of a universally agreed upon conceptualisation of the tourist motivation construct.'

Think point

Make a list of what motivates you to become a tourist. Are there any patterns evident in your list? For example, do some of the items on your list relate to you as an individual, while others are to do with the characteristics of the destination?

Although there is a lack of an agreed common theory to explain motivation, it has nevertheless been a key theme of psychological research. Allusions to why people decide to take certain actions had, up to the beginning of the twentieth century, belonged to the domain of philosophers. One school of thought termed the 'realists' believed that human behaviour was directed by pain avoidance and pleasure seeking, known as the concept of 'hedonism' (Gross, 1992). Alternatively, the 'rationalists' believed that reason determined what people do and therefore people were responsible for their own actions, making a concept of motivation unnecessary.

Early attempts to understand motivation by psychologists were based upon instinct theory, which according to Atkinson *et al.* (1983) represents a very mechanistic view of motivation. Referring to the work of McDougall at the beginning of the twentieth century, Atkinson *et al.* (ibid.) comment that ten inherited instincts were thought to control our behaviour. These were acquisition; construction; curiosity; flight; gregariousness; pugnacity; reproduction; repulsion; self-abasement and self-assertion.

Instinct theory is significant in questioning the assumptions of the early philosophers that man was a rational being, capable of choosing between different courses of action. The theory proposes that innate forces or instincts control behaviour, and shape virtually everything people do, feel and think. According to this theory instincts and behaviour patterns were believed to be inherited. However, many psychologists thought that McDougall's list was too short and by 1924 over 800 separate instincts had been recognised (Gross, 1992). Atkinson *et al.* (1983: 287) comment that: 'Eventually, thousands of instincts had been named, including instincts to estimate the age of each passer-by in the street and to avoid eating apples in one's own orchard.' Thus within this theory a 'tourism instinct' could be added to explain participation in tourism. The main constraint of this approach is that by the labelling of every action as an instinct, little is being explained and individual differences are being ignored.

Probably the most famous psychologist, Sigmund Freud (1856–1939). believed that all human behaviour was driven by two basic instincts: the Thanatos or death instinct, and the Eros or life instinct. Freud developed his theories through working with patients who displayed neuroses, and believed that explanations of individuals' behaviour lay in their subconscious. However, the relevance of the understanding of the subconscious to explain tourism motivation is debatable. Both Dann (1981) and Iso-Ahola (1982) believe that it makes little sense to view leisure and tourism motivation as an unconscious process.

Another theory to explain behaviour is 'drive-reduction theory', which became popular during the 1920s. Atkinson *et al.* (1983: 285) define the concept of a drive as follows:

> A drive is an aroused state that results from some biological need, such as a need for food, water, sex or avoidance of pain. This aroused condition motivates the organism to remedy the need. . . . This is a drive-reduction theory of motivation.

Drive reduction theory has also been applied to explain the seeking of a state of psychological equilibrium beside that of a condition of physiological equilibrium termed 'homeostasis'. According to drive-reduction theory, any psychological imbalance motivates behaviour to restore equilibrium. Emphasis is therefore placed on internal drives to push the person into action to reduce tension.

The concept of psychological homeostasis has specific relevance to tourism studies. According to Fodness (1994), perceived psychological needs and the resulting tension encourage an individual to take action to release the state of anxiety, for example through participating in tourism. Terming this theory as a 'functional approach' to understanding tourism motivation, he comments: 'The extension of functional theory to tourist motivation is straightforward: the reasons people give for their leisure travel behaviour represent the psychological functions (the needs) the vacation serves (satisfies) for the individual' (Fodness, 1994: 560).

Think point

'Drive reduction theory' emphasises that motivations are created by the drive to reduce felt needs, e.g. a need for relaxation. To what extent does participation in tourism help you fulfil psychological or physiological needs, helping you to 'feel better' or return to a more balanced or 'homeostatic' state?

By the 1950s, psychologists began to question the concept of purely innate drives as an explanation of human behaviour, as consideration and recognition was given to the influence of external stimuli upon the individual. Behavioural psychology, emphasising that behaviour is learnt in response to social environments, became prominent. Subsequently, incentives, emotions and cognitions often combine with homeostatic mechanisms to mould basic drives (Davidoff, 1994). For instance, we may become accustomed to expecting pleasure when performing a certain act, such as participating in tourism. Consequently, we learn that being a tourist can give us pleasure, and anticipate it as part of the tourism experience. Expectation may therefore be viewed as an incentive that motivates us to take action.

Similarly, Iso-Ahola (1982) believes that motives are cognitive representations of future states, and therefore we are aware of the reasons for our behaviour. This view is similar to the concept of the 'expectancy-valence' theory of motivation, which states that the direction and intensity of behaviour will be related to expected goals and outcomes that can be achieved through an object, such as tourism (Witt and Wright, 1992). It could therefore be assumed that people participate in tourism with the expectation that it will lead to some kind of reward or rewards.

Think point

To what extent does the expectation of pleasure or having 'fun' while on holiday influence your desire to participate in tourism?

Behavioural psychology also questioned the inherent assumption of drive theory that individuals seek to reduce tension. Instead of seeking to reduce tension, it was noted that some people liked to place themselves in tension-arousing situations. It is possible to hypothesise that participation in certain types of tourism, such as visiting places that are 'unknown' and partaking in certain types of sports with a high risk factor such as mountain climbing, downhill skiing or bungee-jumping, is driven by a desire to place oneself in a tension-arousing situation.

In the view of many psychologists, individuals search for an optimum 'arousal level' and many psychologists have rejected the notions of drive-reduction and homeostasis in favour of this concept (Atkinson *et al.*, 1983). This theory suggests that there is a level of optimal arousal for each individual in terms of the balance between internal and external stimuli. Conditions that depart too severely from this optimal state in either direction incite the organism to restore equilibrium. For instance, too little stimulation or boredom can motivate us to seek stimulating situations, while, alternatively, situations that we find too complex or strange can cause us to withdraw. This is a theory that has been favoured by Iso-Ahola (1980) to explain leisure motivation, as is described in the next section of the chapter.

With the ongoing development of clinical psychology, empirical research led to the development of other theories of human motivation. Just as Freud's psychoanalytical approaches stressing the role of the subconscious had been developed in work with patients, similarly in the 1950s humanistic psychology became more influential primarily through the work of Maslow (1954). He produced a hierarchy of motives that is termed the 'hierarchy of needs'. This model is popular because of its integration of physiological needs with psychological needs; its consideration of wider environmental and social factors; and not least because of its accessibility to a wider audience. A variation of this model related to tourism is shown in Figure 3.2.

Maslow (1954) identified five categories of needs: physiological; safety; love or relationship; esteem; and self-actualisation. Lower-level needs, which Maslow believed to be the strongest, will dominate others higher up in the hierarchy until they are fulfilled. According to Maslow, the individual's future action will be governed by the degree of

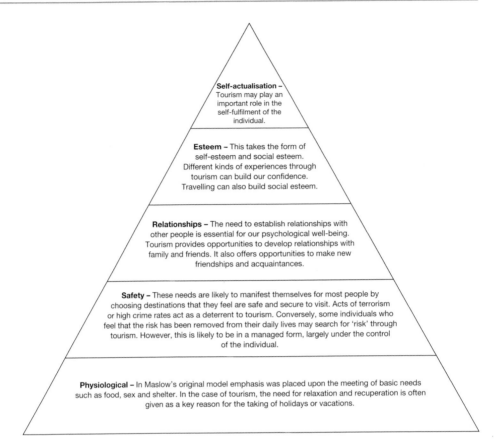

Figure 3.2 Maslow's hierarchy of needs in the context of tourism (after Maslow, 1954).

satisfaction attained within each preceding needs category. Maslow presented the hierarchy as one of a linear progression, the individual moving up to the next level of the hierarchy as they felt their most pressing needs had been satisfied, resulting in them becoming cognitively aware of the desire to satisfy new or 'higher level' needs.

However, a point overlooked in most commentaries on Maslow's seminal work is that to progress to a higher need level in the hierarchy, an individual does not necessarily have to have completely fulfilled the lower-level needs. Witt and Wright (1992: 35) comment: 'He [Maslow] stated that one did not have to satisfy the needs at one level before moving onto the next, and therefore people could be partially satisfied and partially dissatisfied at all levels in the hierarchy at the same time.' The implication of this point is that for each individual there will be different threshold levels in each needs category, so any individual may enter into the next category when they become cognitively aware of the desire to satisfy new needs. Witt and Wright (1992) also point out some of the limitations of Maslow's theory, such as the exclusion of several important needs, including dominance, play and aggression. In addition to this criticism can be added that no time dimension is incorporated in the theory or that no testing of the model in other cultures besides the western one has been carried out.

Think point

Referring to Figure 3.2, what sort of needs have you fulfilled through participating in tourism? How do they relate to Maslow's hierarchy of needs?

The application of motivational theories to tourism studies

As previously stated there is no one commonly agreed theoretical approach to understanding tourism motivation. Fodness (1994: 559) comments in relation to the reasons people give for participation in tourism: 'A widely accepted integrated theory of the needs and personal goals driving these reasons given for travel and the benefits sought from it is however lacking.' Beside the lack of a theoretical perspective, the lack of research into understanding the process of tourist motivation has also been emphasised (Cohen, 1979; Dann, 1981; Fodness, 1994; Mansfield, 1992; Pearce, 1988; Yiannakis and Gibson, 1992).

A seminal empirical study that investigated the motivations for tourism was Dann's (1977) study described in Chapter 2, which emphasised anomie and ego-enhancement. Developing the theme of 'push' and 'pull' factors inherent to Dann's study, Crompton (1979) agrees with Dann on the influence of these factors upon the motivation to travel. He defines push factors as being socio-psychological motives, and based upon 39 in-depth unstructured interviews, Crompton identified the following seven socio-psychological variables as being important in motivating people to travel: escape from a perceived mundane environment; exploration and evaluation of self; relaxation; prestige; regression to a childhood state; enhancement of kinship relationships; and facilitation of social interaction. These are summarised in Figure 3.3.

Think point

Referring to Figure 3.3, can you think of any other types of 'push factors' that encourage people to become tourists?

Dann (1981: 192) also categorises seven approaches to understanding tourism motivation, although these are not exclusive. They are: (i) 'Travel as a response to what is lacking yet desired': the concept of anomie is emphasised; (ii) 'Destination pull': emphasis is placed on understanding the attributes of the destinations that attract the tourist; (iii) 'Motivation as fantasy': emphasis is placed on having free determination to act out one's fantasy role. Dann elaborates, 'Common to this approach is the idea that tourism can liberate tourists from the shackles of their everyday existence' (ibid.); (iv) 'Motivation as a classified purpose': classifications may be made by the purpose of the visit, such as pleasure, novelty and change; (v) 'Motivational typologies': comparison of tourists by either their behaviour or roles; (vi) 'Motivation and tourist experiences':

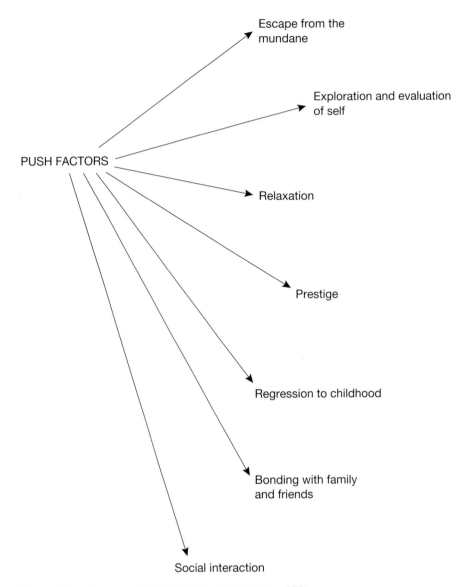

Figure 3.3 Push factors for tourism (after Crompton, 1979).

understanding the motivation of the tourist by interpreting their behaviour; and (vii) 'Motivation as auto-definition and meaning': emphasis is placed on how tourists define situations and particularly how the tourist sees the indigenous or local people.

The diversity of Dann's list led him to suggest that gaining an understanding of the motivation for travel was best attempted through a multi-disciplinary approach including sociology. This approach was subsequently criticised by Iso-Ahola who commented that: 'motivation is purely a psychological concept, not a sociological one' (1982: 257). For Iso-Ahola (ibid.) it is also critical in attempting to understand tourism motivation that the theories that attempt to explain leisure motivation are considered.

The psychological theory favoured by Iso-Ahola to explain leisure motivation is optimal arousal, the basis of the concept being that we search for a level of interaction with our environment that maintains our psychological equilibrium. Applied to tourism this would imply a requirement to be situated in an environment that we perceived as providing the types of activities that provide us with the appropriate level of optimal arousal. Our choice is likely to be influenced by a reflection of the level of stimulation in our home environment. Consequently, if we feel over-stimulated, perhaps experienced in the form of stress, we would be likely to search for a destination in which we perceive we could find tranquillity and relaxation. Conversely, if we feel under-stimulated at home, we may search for a higher level of stimulation. This may manifest itself in a desire for high risk sports-based activities such as downhill skiing or all-night raves. There is also a third type of possible relationship in which we feel neither under- nor over-stimulated at home, subsequently the desire for optimal arousal is not a great influence on our holiday choice. These possible relationships are shown in Figure 3.4.

Key components of the intrinsic motivation to move towards an optimal level of arousal are, according to Iso-Ahola (1980), the needs for competence and self-determination, alternatively labelled as 'perceived freedom'. As he suggests, needs are

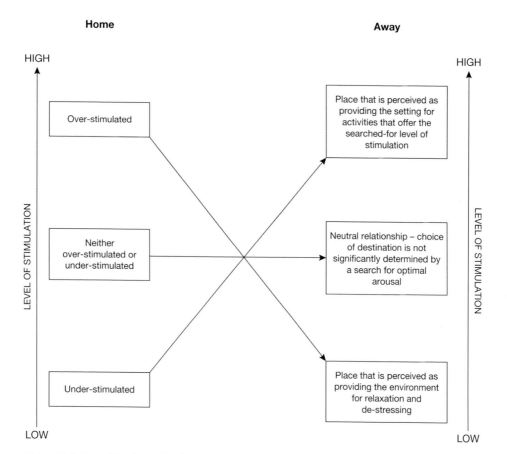

Figure 3.4 Searching for optimal arousal through tourism.

the most readily observable part of behaviour. For instance when a person is asked why they participate in a recreational activity, they do not state 'optimal arousal', but do feel more easily able to identify their needs such as freedom and relaxation. Iso-Ahola suggests that perceptions of freedom and competence are determined by comparison to others and by past situations. He also adds that human actions are motivated by a cognitive awareness, resulting in subjectively designed goals and perceived rewards, for example that participation in tourism will produce an end-state of satisfaction.

The psychological theory of seeking optimal arousal as an explanation of leisure behaviour has similarities to the concept of 'flow', described by Csikszentmihalyi (1975). He explains 'flow' as a subjective experience that is achieved where the challenges of a situation are matched by a person's skills. There would therefore seem to be a corollary to the concept of 'optimal arousal'. He identifies seven key aspects of the flow experience: (i) the perception that challenges and personal skills produced by an activity are in the balance; (ii) the centring of attention; (iii) loss of self-consciousness; (iv) an unambiguous feedback to a person's actions; (v) feeling of control over actions and environment; (vi) a momentary loss of anxiety and constraint; and (vii) feelings of enjoyment or pleasure.

When in a state of flow there is no space in the consciousness for distracting thoughts and self-consciousness disappears. Csikszentmihalyi (1997: 28) describes flow as: 'The metaphor of "flow" is one that many people have used to describe the sense of effortless action they feel in moments that stand out as the best in their lives.' He suggests that there are two main 'flow channels', either side of which lies anxiety (where challenges are perceived as being greater than skills), and boredom (where skills are greater than channels). To date, the ideas of Csikszentmihalyi have had little or no application directly to aiding the understanding of participation in tourism.

For certain types of tourism, such as adventure holidays, the concept of 'optimal arousal' and 'flow' may be central to understanding the motivation for travel. The importance of arousal, expressed in the form of risk-related elements, e.g. challenge, adventure and excitement, as a satisfactory experience for adventure holidays, was assessed by Johnston (1992) in a study of 915 individuals participating in various types of mountain tourism in New Zealand. For 32 per cent of the sample, risk increased the enjoyment of the recreation vis-à-vis 23 per cent for whom it did not. Interestingly, for 38 per cent of respondents, risk could both increase and decrease the level of enjoyment. The three most common definitions of risk were as a 'challenge'; 'danger'; and 'uncertain outcome'. A key point of Johnston's results was that although excitement and challenge were essential ingredients for a satisfactory experience, there was a need for 'managed' risk levels, to make sure the participant did not feel or perceive themselves to be in too much danger.

Think point

Do you search for 'stimulation' through tourism? How does this manifest itself, e.g. intellectually, sexually, a search for danger or thrills? Alternatively, do you view tourism as a time to escape from the stress of life at home and a time to relax? How does your home environment influence the choice of where you go on holiday or vacation?

In terms of understanding the satisfaction that may be achieved by an individual from participation in tourism, key components are the themes of intrinsic as opposed to extrinsic motivation (Iso-Ahola, 1980; 1982). Both intrinsic and extrinsic motivation influence goal setting and the establishment of criteria against which the level of satisfaction with the experience will be judged. When an activity is performed to obtain a reward that is external to the activity, it is said to be extrinsically motivated. When no apparent extrinsic reward is present, one's behaviour is said to be intrinsic. For example, we may choose our employment predominantly on the grounds of the extrinsic reward of a high salary, or on intrinsic rewards such as helping others or self-autonomy.

The influence of extrinsic vis-á-vis intrinsic motivation upon the level of satisfaction gained from participation in an activity is demonstrated by research conducted by Fielding *et al.* (1995). Based upon a sample of 187 tourists who set out to climb Ayres Rock in Australia, a distinction was made between tourists who were climbing the rock for the sake of it, or 'intrinsically motivated', and those tourists who were climbing the rock solely for the purpose of reaching the summit or externally motivated by achievement. The results showed that 82.7 per cent of intrinsically motivated climbers reported high levels of enjoyment, whereas only 61.5 per cent of achievement-motivated individuals reported this level of enjoyment. Conversely, 73.5 per cent of those who reported low levels of enjoyment were achievement motivated, the remainder being intrinsically motivated.

Think point

How could 'extrinsic' and 'intrinsic' motivations be expressed in terms of taking holidays? Think of the reasons or motivation why people take holidays and divide them into two respective groups.

The travel career

A further consideration of the motivation for travel and tourism is how it evolves and changes with time and experience. A theory of tourist motivation incorporating a longitudinal dimension and multivariate analysis is proposed by Pearce (1988), who suggests that people have 'careers' in tourism, just as at work. Similar to a work career, Pearce argues that a travel career is both consciously determined and purposeful. However, an important difference is that given that participation in tourism is mostly out of choice, our travel career is more likely to be weighted towards intrinsic rather than extrinsic motivation. The basis of the 'travel career' is that motivations for participating in tourism are dynamic and will change with age, life-cycle stages, past experiences of tourism and the influence of other people.

Pearce (1988) tested the concept in theme-park settings in Australia, and after segmenting the market by demographic groups, demonstrated that differences existed between the levels of importance attached to different categories of needs by these groups. For example, 13–16-year-olds placed a much higher level of importance on 'thrills' than did family groups. The main components of Pearce's multivariate model are five cate-gories of needs: relaxation, stimulation, relationship, self-esteem and fulfilment, which are similar to Maslow's (1954) hierarchy of needs. Pearce envisages a progression through this hierarchy of needs, up the 'leisure ladder', as is shown in Figure 3.5.

Although the concept has been criticised by Ryan (1997) as being too simplistic in its interpretation of general tourist behaviour, accepting that motivation in leisure and tourism is multivariate, it is possible to adapt Pearce's construct to a variety of recre-ational settings. It would also be possible to assess how these needs change for an indi-vidual using a longitudinal perspective incorporating life-cycle stages and tourism experiences.

Pearce's work emphasises 'needs analysis' as a way of interpreting motivation. A theme of motivational research is that expressed needs represent a cognitive awareness

---

Think point

Referring to the 'leisure ladder model' shown in Figure 3.5, think of examples of how the needs in each category can be fulfilled through tourism. How might the emphasis placed upon the needs change with age, life cycle, and past experiences of tourism?

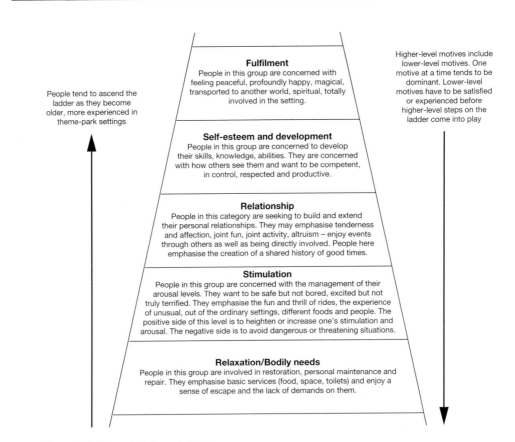

Figure 3.5 Pearce's leisure ladder.

of the future benefits from taking a certain course of action, in this case participation in tourism. The work of other researchers, such as Crompton (1979) and Crandall (1980), suggests that the use of needs to explain participation in leisure and tourism is both multivariate and complex. Through the experience of tourism, which involves travelling from one's home environment, and interacting with unfamiliar cultural and physical environments, a complex range of needs is likely to be fulfilled.

Personality

For many psychologists another key determinant of behaviour besides that of motivation is personality. Yet, although personality is perhaps one of the best-known topics associated with psychology, similar to motivation there exists no definitive definition within the discipline as to what it actually is. In attempting to define personality, Decrop (1999: 125) defines it as: 'a reflection of a person's enduring and unique characteristics that urge him or her to respond in persistent ways to recurring environmental stimuli.' Nevertheless, although there may be a lack of agreement upon what constitutes a personality, this does not imply that personality theories are not useful in attempting to predict behaviour (Ross, 1998).

In terms of the application of personality theory as a predictor of travel behaviour, notable work has been undertaken by Plog (1973, 1994). His works are based upon the concept of psychographics, which has its base in psychoanalysis. Its application to tourism involves examining and trying to comprehend a tourist's intrinsic desire to choose a particular type of holiday or destination, by measuring their personality dimensions. In terms of aiding the understanding of intrinsic needs, Plog (1994: 212) comments that psychographic research allows the researcher 'to get inside the skin of the traveller', to understand why people select special places to visit, and pursue specialist activities at certain destinations.

Plog's contention is that socio-demographic characteristics such as social class, age and gender have decreasingly less use in terms of predicting both leisure and tourism behaviour in postmodern societies. For example, what were historically elitist activities, such as travel to the French Riviera or Caribbean, or participating in downhill skiing or golf, are now available to a wider social spectrum of society than in the past. Consequently, attributing certain types of leisure and tourism behaviour to certain socio-demographic groups is increasingly less certain than in the past. The method proposed by Plog is to identify group typologies, based on people who have shared motivations and attitudes, as a means of identifying travel patterns.

Plog's (1973) initial research was developed in a commercial consultancy setting, working with airline business clients in the US who were concerned with understanding the personality characteristics of people who were flyers as opposed to non-flyers. Using telephone interviews to ask questions relating to personality types and travel patterns, Plog subsequently established typologies of tourists, ranging on a continuum from what he termed psychocentrics to allocentrics. The psychocentrics were found to display characteristics of being self-inhibited, nervous and non-adventuresome people, while allocentrics displayed characteristics of being variety seeking, adventurous and

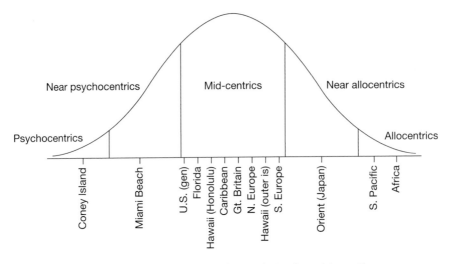

Figure 3.6 Plog's original model of the psychocentric to allocentric continuum.

confident people. Subsequently, allocentrics were more likely to be flyers than non-flyers. The majority of people belonged to the mid-centric groups, displaying dominant characteristics of neither psychocentrism nor allocentrism. Plog's (1973) model is shown in Figure 3.6, displaying the typical destinations that would be chosen for each psychographic typology.

Think point

Plog's (1973) original model shown in Figure 3.6 is now over 30 years old and was developed specifically for the North American market. Redraw the model, showing the typical destinations that different psychographic groups from your community would be likely to travel to.

Plog (1973) also established a relationship between the type of traveller and how destinations develop in terms of their popularity. He suggested that based upon their personality characteristics, new tourism destinations are likely to be discovered by allocentrics. As the allocentrics talk to friends about their experiences, the destination becomes more popular and becomes frequented by a 'near-allocentric' group. A larger group in numbers than the allocentric group, their arrival encourages the development of tourist facilities including hotels and restaurants. The diffusion of information about the destination and an increasing familiarity with it, leads to the attraction of mid-centric tourists, leading to the development of the destination into a 'mature' stage. By this stage, the allocentrics have ceased to visit the destination, as the naturalness that attracted them in the first place has been lost. At this stage, according to Plog (1973: 58): 'The destination has reached its maximum potential because it is now attracting the broadest audience possible.' In Plog's view the destination will continue to progress in terms of its appeal towards the psychocentric end of the continuum, although he does not fully explain why, apart from hinting at over-development and loss of qualities which attracted the allocentrics.

The ideas of Plog (1973) on destination development can be associated with Butler's (1980) theory of the destination life cycle, which is explained in Chapter 7.

Attitudes and environmental psychology

While motivations and personality are important in understanding tourist behaviour, another key aspect from a psychological perspective of this complex jigsaw is attitudes. In trying to understand tourist behaviour, the attitudes held by tourists to the environment of a destination are likely to influence the activities they decide to undertake in it. Attempting to explain what an attitude is, Dibb *et al.* (1994: 115) comment: 'Attitude refers to knowledge and positive or negative feelings about an object or activity. The objects or acts towards which we have attitudes may be tangible or intangible, living or non-living.' According to Malim and Birch (1992) attitudes have three components: the cognitive, the affective and the behavioural. The cognitive component is concerned with perceptions or beliefs, for example 'I think Greece possesses the most wonderful historical sites in the world.' The affective response involves the feelings or emotions generated by the object, in this case the historical site: 'I like historical sites very much.' The behavioural or conative response relates to the behavioural intention, based upon the cognitive and affective responses, e.g. in this case 'I will choose Greece for my vacation.' Attitudes therefore involve a cognitive dimension of information processing. As Decrop (1999: 103) suggests: 'Tourist perception can be defined as the process of translating tourist information from the external world into the internal, mental world that each of us experiences.'

Attitudes are therefore fairly permanent sets of evaluations, which we carry around with us and affect how we interpret people and things, making us more predictable in our responses than we might otherwise be. However, this does not mean that they cannot be challenged or changed. Given that they are based upon a cognitive component and that we have the ability to assimilate new information, the cognitive part of an attitude is susceptible to influence, with subsequent changes to the affective and behavioural components. Thus, for example, in terms of the marketing of destinations it is possible to present information about destinations in a way that may overcome unfavourable cognitions or beliefs about those places. Similarly, it is possible to change racist or homophobic attitudes that may act as barriers to travel for minority groups in society, or tourist behaviour that impacts negatively upon nature.

Think point

Choose five tourism destinations that are popular for tourists from your own country. For each of them, identify your own cognitive, affective and behavioural responses to them.

Linked to the attitudes that tourists have to destinations is an area of psychology that has developed during the latter part of the twentieth century; that of environmental psychology. If we are to understand tourism more comprehensively, it can be argued that comprehending how tourists perceive and experience foreign environments represents an essential component of this understanding. Ittleson *et al.* (1976) distinguished five different, but overlapping, modes of interaction that we may have in experiencing our physical environment, as shown in Figure 3.7.

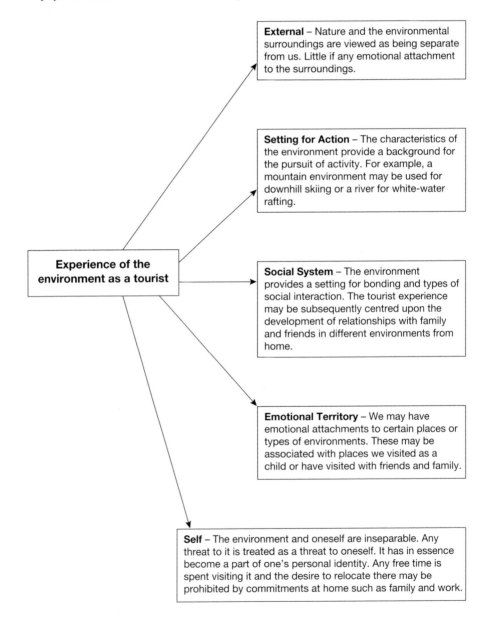

Figure 3.7 Modes of environmental experience in tourism (after Ittleson *et al.*, 1976; Iso-Ahola, 1980).

The modes represent a range of feelings about our surroundings, from being external to us to one where they are internalised. However, they are not exclusive, with one environment being experienced in two or three different modes at the same time, e.g. a mountain trekker may experience the environment as a 'setting for action', a 'social system' and as 'self', within the same visit. How we view a particular environment is likely to determine how we interact with it as a tourist. If we perceive it as being external to ourselves there is an increased likelihood we may use it in an uncaring way. This may manifest itself in a variety of types of behaviour, from the shooting of wildlife to throwing litter away carelessly. Conversely, if we view the environment as 'self', it is likely we would be willing to take action to protect the environment from any use that we viewed as being detrimental to its quality.

The experiencing of the environment as 'external physical space' in the context of tourism can be associated with its use as a 'setting for action'. The view of the environment as 'external space' is developed around the ideas of Proshansky (1973) who proposes that capitalism encourages people to view the environment as an external entity, where there is a clear demarcation between oneself and the natural environment. How such a relationship translates into tourism is untested but it is suggested that it is likely that a tourist will either be involved in an activity that demonstrates a wish to conquer nature or treat it in an instrumental fashion to meet their needs. For example, the former mode is encapsulated in the following description of the experience of a skier in the early twentieth century:

> I glory in the victory over self and Nature . . . the greatest of all joys of skiing is the sense of limitless speed, the unfettered rush through the air at breakneck speed. . . . Man is alone, gloriously alone against the intimate universe. . . . He alone is Man, for whose enjoyment and use Nature exists.
>
> (Smout, 1990: 14)

Further examples of how the environment may be used as a setting for action are shown in Box 3.1.

A third type of interaction is the viewing of the environment as a 'social system'. Emphasis is placed upon use of the environment as a social setting and establishing and furthering social relationships. Different types of environments are likely to be viewed as being more or less conducive for social interaction. For instance, in the work of Pearce (1993) already referred to, social interaction was found to be an important part of the tourism experience. In contrast Iso-Ahola (1980) suggests that the more naturally an environment is viewed, the less one expects to experience it as a social system, instead emphasising the experience of solitude. He also makes the point that experiencing the outdoor environment as a social system can be internally or externally induced. For example, an outdoor environment may be visited because there are few perceived alternatives, and the experience becomes that of a social system because of the presence of others at the location, thus the situation is externally induced. Alternatively, an outdoor environment may be visited specifically to spend time with family or friends, which is internally induced. If the experience of the environment as a social system is internally induced, it is likely to be rewarding to the individual; if it is externally induced, the level of individual reward is likely to be lower.

Box 3.1 The use of the environment as a 'setting for action'

The perception of the environment as a setting for action is encapsulated in the following passage of an advertisement for a holiday in Tenerife from the *Summer '99: Have it your way* brochure:

> We all piled into Bobby's the biggest club in Americas, where they had this brilliant English band and a rather (ahem) adult comedian – certainly broadened our knowledge if not our culture. The reps were all there too and really getting into it. Fuelled by that, we exchanged girls with hip strings for whales with blow holes on Atlantic Knights. What a mad day . . . a Freestyle cruise out to sea to spot dolphins and whales and other aquatic things. Oh and to drink a large amount of free beer.
>
> (Club Freestyle, 1999: 32)

A further example of how the environment may be used as a setting for action is given by Drumm (1995: 2) commenting on backpacker operations that take visitors into the rainforest in Ecuador:

> Only 20% [*sic*] of local guides have completed secondary education, and very few are proficient in a language other than Spanish. Together with a social context which embues them with a settler frame of mind, antagonistic to the natural environment, the negative impacts of this operation are significant. Hunting wild species for food and bravado is commonplace during tours as well as the occasional dynamiting of rivers for fish. The capture and trade in wildlife species including especially monkeys and macaws is also very common.

The fourth type of experience is of the environment as 'emotional territory'. As the terminology suggests in this case the environment may be experienced emotionally. This emotional link through tourism to the environment would usually be experienced in a positive way. For example, the inspiration of the Romantic movement of the nineteenth century rested in the emotionality and spirituality of nature. Visiting natural ecosystems such as coral reefs, mountain areas, or built environments such as Venice may similarly inspire a sense of emotional awe in the environment. Emotional links may also be experienced through the association of a particular place with childhood memories; subsequently they are revisited with one's own children, maintaining an emotional link between place and person through the generations. However, as Iso-Ahola (1980) points out, continued exposure to and familiarity with an environment may reduce emotional feelings.

At the extreme of emotional attachment to a particular place, an environment may become interwoven with one's own identity, being viewed as an inseparable part of oneself. Consequently, any damage or harm that occurs to the environment transcends into hurt to oneself (Iso-Ahola, ibid.). This concept is also analogous to Cohen's idea of the 'existential' type of tourist, explained in Chapter 6, who has an 'elective centre' of where they really belong away from where they are forced to live. A person may thus feel their spiritual and emotional centre lies in another place from where they actually live, but they are prohibited from relocating there because of obligations such as work or family. The migration of people from northern Europe upon retirement to places in the warmer environments of southern Europe that they have frequented as tourists, and people who own holiday homes and wish to relocate there, but are prohibited from doing so for economic reasons, are testimonies to this type of feeling.

Similar to Pearce's idea of a career in tourism, Iso-Ahola (1980) suggests that how we interact and emotionally associate ourself with a place is likely to change with one's life span and past experiences. There is therefore an implied dynamic element of changing environmental experiences across the life span. As Ittleson *et al.* (1976: 206) put it: 'Environmental experience is the continuing product of an active endeavour by the individual to create for himself a situation within which he can optimally function and achieve his own particular need for satisfaction.'

Think point

How have you experienced environments in different ways as a tourist? For example, have they provided a background for fun and action; or have they been used primarily to develop family bonds or friendships? Is there any particular place away from your home that has evoked emotions because of its characteristics and associations? Is there a place that you identify with as 'a part of you', in the sense that if it were in danger of being changed or spoilt from development or pollution, you would feel upset?

The visitor experience

As was commented upon at the beginning of the chapter, some of the earliest applications of psychology to tourism were associated with observations of visitors to museums. More contemporarily, psychological theories are being used to enhance the visitor experience. Included within this are elements such as helping the visitor at a destination or site find their way around and providing environmental interpretation of what they are seeing.

Cognitive mapping is concerned with understanding how we learn about spatial information and how we learn to move around our environment (Moscardo, 1999). Although research into the cognitive maps that tourists have of areas they have visited is a neglected area, such research offers the potential to aid the understanding of how individuals come to assimilate the new and the unfamiliar. An understanding of how tourists become acquainted with the characteristics of destinations also has implications for their promotion and commercial viability (Ross, 1998).

Moscardo (1999) notes that children and new arrivals to an area seem to build up their knowledge of a place by first identifying and remembering distinctive features or landmarks, referred to as the 'anchor point' theory. In a review of existing research from environmental psychology into cognitive maps, Moscardo (ibid.) found that many map users do not use compass points or compass directions to find their way around, nor do they draw correct angles of turns and bends. Subsequently, it is suggested that stylised maps such as route maps may be more useful for visitors to tourist sites.

Besides helping visitors to find their way around new places, the visitor experience can also be enhanced through interpreting the environmental setting. As Ross (1998) suggests, an important aspect of environmental interpretation is communication. Moscardo (1999) suggests that there is a need to develop 'mindfulness' in a visitor vis-à-vis 'mindlessness', as when we are mindful we are paying attention to what is around us. Subsequently, in a state of mindfulness we are open to learning about the environment. The conditions to encourage mindfulness include new and different settings; varied and changing situations; control and choice; and personal relevance (Moscardo, ibid.). He also suggests that the ten most important features of communication or interpretation of an activity are: it arouses interest in the subject; the information is clearly presented; it teaches something new; you can't help but notice it; it gets the message across quickly; it involves the visitor; the visitor can take it at his own pace; it is a memorable exhibit; it respects the intelligence of the visitor; and it uses familiar things or experiences to make the point.

Alongside arousing the interest of the visitor and enhancing their experience, environmental interpretation and communication may also be effective in changing any aspects of negative tourist behaviour towards the surroundings (Bramwell and Lane, 1993; Mason, 2003). By having an impact on the cognitive component of attitude formation, it is possible to develop more respectful and positive emotive responses from tourists to their surroundings, which would be likely to encourage more respectful behaviour.

Summary

- Psychology is of special relevance in aiding the understanding of the motivations involved when people choose to become tourists. Socio-psychological motives include the need for escape; self-exploration and evaluation; relaxation; prestige; regression to childhood; and social relationships. The needs that are fulfilled through tourism are likely to change with age, life cycle and past tourist experiences, hence we can be viewed to have a 'travel career' just as we might have a work career.

- Personality is also an important factor in attempting to predict tourist behaviour. Personality type may in the future prove to be a more reliable predictor of tourist behaviour than socio-demographic characteristics such as age, gender or class. Attitudes also determine our behaviour as tourists, including the choice of where we travel. Attitudes are composed of cognitive, behavioural and affective dimensions.

- Environmental psychology can help in understanding how we view the space that we inhabit as tourists. There are a range of experiences we may have as a tourist ranging from experiencing the environment as an 'external' physical space to experiencing it as 'self'. All are likely to influence our behaviour and impact as a tourist.

Suggested reading

Maslow, A.H. (1954) *Motivation and Personality*, Harper, New York.

Pearce, P.L. (1988) *The Ulysses Factor: Evaluating Visitors in Tourist Settings*, Springer-Verlag, New York.

Ross, G.F. (1998) *The Psychology of Tourism*, 2nd edn, Hospitality Press, Elsternwick, Victoria.

ECONOMICS AND TOURISM

4

This chapter will:

- consider how economic theory can be applied to tourism;
- explore the relationship between microeconomics and tourism;
- critically discuss the macroeconomic benefits of tourism;
- consider tourism as a form of trade.

Introduction

The increase in demand for tourism, explained in Chapter 1, has subsequently led to it having a global economic significance, as is exemplified in the following sentence from the World Travel and Tourism Council (2002: 3):

> According to WTTC research, Travel and Tourism generates economic activity worldwide representing over ten per cent of total global GDP. The industry accounts for over 200 million jobs (direct and indirect). With 4.5 per cent growth forecast per annum for the next ten years, Travel and Tourism is not only one of the world's largest, but one of its fastest growing industries.

For tourism to make this level of contribution to the global economy relies upon individuals, households, firms and governments making decisions to allocate resources to tourism. The social science discipline that is particularly concerned with understanding how individuals and society allocate resources is economics.

What is economics?

Definitions of the meaning of economics emphasise it as a study of how resources are allocated, as exemplified in the following definitions from Begg et al. (2003: 4): 'Economics is the study of how society decides what, how, and for whom to produce'; and Mankils (2001: 4): 'Economics is the study of how society manages its scarce resources.' However, Lundberg et al. suggest that scarcity is a human phenomenon; a consequence of human needs and wants. They subsequently define economics as: 'a social science that seeks to understand the choices people make in using their resources to meet their wants' (1995: 27). Accordingly, economics is concerned with three basic economic questions: what goods and services to produce; how to produce them; and for whom in society are the goods and services to be produced (McLeish, 1993).

The scarcity of available resources to meet all human needs and wants means that society cannot produce all the goods and services people may wish to have. As Baumol and Blinder (1999) comment, despite the dramatic improvement in standards of living since the Industrial Revolution, human society has not reached a state of unlimited abundance, so people are forced to make choices. Given that we can't have everything that we want, people must trade-off one goal against another, and make choices about which goods and services they wish to consume. The requirement to choose one good or service in preference to another is termed an 'opportunity cost', in reference to the opportunity we have had to relinquish to have what we want. For example, to go on holiday may involve the opportunity cost of forgoing the spending of money on the next-best alternative, such as a hi-fi system or computer. Opportunity costs also occur at a national level, as governments, like households, have limited budgets, and therefore they have to decide to prioritise certain sectors of the economy for expenditure in preference to others (Tribe, 1999). For example, should governments allocate resources to promote their country as a tourism destination and support infrastructure development for tourism, or should they allocate the same resources to improve schools and hospitals?

Think point

Few people possess an abundance of resources, usually measured in terms of money, to have everything they want. Therefore, we often have to choose to purchase one item in preference to another, termed by economists as an 'opportunity cost'. Have you ever made the decision to spend money on a holiday or travelling in preference to purchasing something else? If so, for what reasons?

A major concern for economists in finding the best mechanism for society to allocate resources is to maximise 'efficiency'. Expressed in economic terms, efficiency can be explained as meaning that society is getting the most of what it can from its scarce resources. The choice of how best to allocate society's resources falls between two main mechanisms. One is the 'command economy', in which faith is placed in central planners in the government as being in the best position to guide economic activity. Subsequently in this system, government decides which goods and services to produce, their quantity, and who should produce and consume them. Emphasis is placed on the government as being the only entity capable of organising economic activity in a way that promotes economic well-being for the whole of society. This system was the one favoured by the communist countries of Eastern Europe until their collapse early in the last decade of the twentieth century. According to Mankils (2001) the collapse of communism may be the most important change in the world during the last half-century, there now seeming to be no other choice to the second main mechanism for resource allocation, the market system.

Think point

We are used to purchasing travel and tourism, our choice of holiday or vacation usually being determined by our ability to pay for it. In a command economy in which the government controls the tourism industry, what kind of alternative priorities and criteria could be used to decide who should participate in tourism, other than purely the ability to pay for it?

In a market economy, the decisions about the allocation of resources are made by millions of individual firms and households, rather than by central planners. The basis of this interaction is that firms decide what to produce in response to the demands of the households. The interaction between the firms and households takes place in the market, where prices and self-interest guide the decisions of consumers and producers, thereby regulating demand and supply. Today, the market mechanism is commonly held to be the best mechanism for the allocation of resources for the production of goods and services. However, there are some services for which there seems to be no market, notably national defence and the judiciary. Consequently, many economies are mixed economies with resource allocation often being decided by governments for essential services including national defence, law, health and education, and by markets for other goods and services.

The thesis of markets as the best mechanism for allocating resources can be traced to over 200 years ago and the seminal work of Adam Smith (1723–1790). Previously referred to in Chapter 1 as probably the most eminent tutor on the Grand Tour, Smith published his most renowned work *An Inquiry into the Nature and Causes of the Wealth of Nations* in 1776, just before the Industrial Revolution was about to commence. Smith's observations were based upon changes in society and economy that had been occurring during the eighteenth century, including expanding domestic and international trade and commerce; improvements in agriculture; population growth; and the establishment of complex economic institutions like banking and the crediting system (Gill, 1967). Smith made probably the most famous observation in economics, that households and firms interact in markets as if they are being guided by an '*invisible hand*', which leads them to desirable market outcomes for both themselves and society. The basis of this concept is that through intentionally serving one's own interests, one unintentionally serves the interests of society as a whole. Based upon this principle, Smith advocated the free-market economy as the best allocator of resources, as is explained in Box 4.1. According to Smith the advantages of markets are twofold: (i) they ensure that producers supply the commodities that are demanded by consumers at a price which represents their 'worth'; and (ii) they guarantee good management of resources to keep production costs low (Gill, ibid.). Smith advocated that individuals are usually best left to their own devices with minimum state interference; a general philosophy of economic life referred to as the doctrine of 'laissez-faire' or the 'system of natural liberty'.

> ## Box 4.1 'The invisible hand'
>
> In 1776, Adam Smith (1723–1790) published one of the most famous works of economics, *An Inquiry into the Nature and Causes of the Wealth of Nations*. In terms of allocating resources for the benefit of society, Smith emphasised that nature is man's best guide. That is, God has arranged things so that through pursuing our own self-interest, the working of Providence's 'invisible hand', will ensure we naturally act for the best of society. The basic premise being that by enriching themselves, people also enrich society. The spur for self-interest is the profit motive, achieved by producing goods that are cheaper for people to buy than to produce for them-selves. If anyone becomes too greedy and tries to make too much profit at the expense of society, other producers will compete with them to lower prices. In this way markets regulate themselves for the benefit of society. Consequently, Smith argues, since the invisible hand works so well, governments should not interfere with produc-tion and trade. The principle of Smith's theory remains the cornerstone of contem-porary neo-classical economics.

Microeconomics and tourism

The branch of economics that is concerned with the behaviour of consumers and firms, and the determination of market prices, is referred to as 'microeconomics', making it distinctive from the workings of the whole economy or 'macroeconomics' which is dis-cussed in the next section of the chapter. The demand for tourism and consumer behav-iour is likely to be influenced by a range of factors, including levels of disposable income; price; comparative quality; fashion; advertising; time free from obligations such as work; and demographics (after Tribe, 1999). The interlinkage of these factors is exemplified in Box 4.2 concerning the decline in tourism demand for the Cote d'Azur in France.

The case of the Cote d'Azur illustrates the complexity of factors that determine demand for tourism destinations besides that of price. However, when we consider the demand for many of the services supplied in the tourism market, price would seem to be an important determinant. Notably, as was discussed in Chapter 1, mass participa-tion in both domestic and international tourism was, in part, a consequence of the inter-vention of entrepreneurs in the market and a subsequent lowering of the price of travel. More contemporarily, the popularity of 'budget' airlines; the supply of discounted stays at hotels; and discounted holiday packages all indicate that price seems to influence demand for tourism.

Think point

Typical factors that influence consumer demand for tourism include levels of disposable income; price; comparative quality; fashion; advertising; time free from obligations such as work; and demographics. How influential are these factors upon your decision-making of where you go on holiday or vacation?

Box 4.2 Demand for tourism to the Côte d'Azur in the south of France

The Côte d'Azur in the south of France has been one of the most popular regions for tourism in the world since the nineteenth century and has a long association with artists, film stars and millionaires. However, in 2004 visitors were expected to be 20 per cent less than in 2003, and between 2000 and 2003 there had been small but steady decreases in the numbers of visitors each year. The reason for this, according to the director of the tourist office in Antibes, is that the Côte d'Azur has 'gone out of fashion'. Sage (2004: 35) cites Baute as commenting:

> The truth is that a holiday here is no longer a must. A decade ago, you just had to come here if you wanted to be seen as trendy. Today, people go to Croatia, or North Africa, or Cuba, or somewhere like that. Or they look at their bank accounts and decide they're not going anywhere.

The president of the Union of Hotel and Restaurant Owners in Antibes and Juan-les-Pins comments: 'The generation that made the Riviera was aged between 30 and 40 in the 1950s', while a local entrepreneur in reference to the market comments: 'They've died or when still alive they're in retirement homes. The younger generation just don't come here.' Riviera hoteliers also blame the introduction of the Euro for their problems, saying it has led to increased inflation and left families with less disposable income. The introduction of the Euro has also had the effect of allowing tourists to directly compare prices for different tourism goods and services between European countries, permitting them to see how expensive the Riviera is.

Source: Based upon Sage (2004)

A key concept in assessing the relationship between change in demand and change in price is the theory of 'price elasticity'. When demand is 'inelastic' it is less responsive to a change in price, where: 'a change in price produces a less than proportionate change in the quantity demanded' (ibid.). By contrast, 'elastic demand' is more sensitive to price change, where: 'a change in price results in a more than proportionate change in quantity demanded' (Pass *et al.*, op.cit.). In the case of travel and tourism, different types of tourism are likely to have different degrees of elasticity. For example, the budget airlines and many large tour operators have been successful because of the managing of supply to guarantee low prices, and subsequently operate in a price elastic market. In some cases the demand for tourism will be considerably more price inelastic. For example, the demand for business-class air travel is as much a function of the quality and convenience of the service as its price, as are famous hotels, or other tourism products placed at the more luxury end of the tourism market.

A 'high demand and low price' strategy is central to the success of mass-market tour operators, who play a highly significant role in some tourism markets. For example, the German, British and Dutch markets account for approximately two-thirds of the package tours sold in Europe, with approximately 18 million being sold in Germany and 15 million in Britain per annum. A consequence of this high demand and low-price model has been the extension of the holiday season in many destinations, and a diversification of the demographic profiles of holidaymakers to include all age groups and socio-economic sectors of society (Laws, 1999).

However, the competitiveness of a market based primarily upon price makes companies particularly vulnerable to market downturns. For example, the rapid growth in the European budget or 'low-cost' airline market since the 1990s has received copious praise as a model of innovative entrepreneurship. However, in the first decade of the twenty-first century, many firms are struggling to compete in this marginally priced market. Ryanair, the longest-established budget airline and one of the main European players, recorded its first ever loss in ten years of operation in the first three months, or quarter, of 2004, while other low-cost airlines including Planet Air, SkyNet, Jet Green, Jet Magic, Good Jet, Ciao and Flying Finn have all ceased trading (Clark, 2004). Oversupply in the market has become a problem, as Clark (2004: 12) suggests: 'There are too many seats available in too many aircraft for rock-bottom prices which are simply too low to be profitable.'

Think point

Compare advertisements for different types of tourism products, e.g. package holidays; budget air travel; business-class air travel; hotels; and ecotourism. What kinds of factors are emphasised to try to persuade a consumer to purchase the product? How important is price as a selling point between the different types of tourism products?

Market failure

However, the market will not necessarily always allocate resources for tourism or other forms of human activity efficiently, in the sense of providing benefits for 'all of society'. For example, while the advent of mass tourism has resulted in many benefits in terms of widening participation in international tourism and providing economic benefits for destinations, it has also been found to have a range of negative impacts upon sections of destination communities not directly involved in the tourism market, and also upon nature. Where parties external to a transaction have their well-being affected by it, these effects are termed 'externalities' (Baumol and Blinder, 1999), and they represent one type of 'market failure'. Though externalities may be positive, they are frequently negative, and it is these that threaten the well-being of society. Examples of the types of negative externalities that may occur from tourism include various types of pollution, increased crime rates and traffic congestion.

Another reason for market failure is the creation of a monopoly or oligopoly, i.e. the ability of a single person or a small group of people to unduly influence market prices. As Ellwood (2001) points out, Adam Smith was adamant that markets would work most efficiently in allocating resources when there is equality between buyers and sellers, with neither the buyer nor seller being large enough to control the market price. At the end of the 1990s, the European Commission raised concerns of an oligopoly arising in the tourism industry, creating a situation where major tour operators would have an undue influence upon market prices. Particular concerns rested with the £850 million bid by Airtours, the second largest package-holiday group in Britain, to takeover its rival First Choice, who was the fourth biggest. As Buckingham and Wolf (1999: 8) comment: 'The European Union's competition regulator said it was concerned that the acquisition would give the remaining three biggest travel companies – Thomson, the merged Airtours-First Choice and Thomas Cook – too much clout in the British and Irish markets.' These three companies would be selling 75 per cent of the foreign package holidays in Britain and the Commission ultimately blocked the deal. One year later, another proposed merger was subject to the Monopolies Commission, this time the proposed £1.8 billion takeover of Thomson Holidays by the German company Preussag, in an attempt to create the world's largest travel company. This time the deal was approved after Preussag agreed to sell its stake in Thomas Cook to Carlson of the US.

In a wider societal context, the most dramatic example of market failure was the 'Great Depression' of the 1930s, which acted as a catalyst to the development of what is referred to as 'macroeconomics'. The Great Depression was the time of a worldwide slump in economic output and prices occurring between 1929 and 1934, the catalyst being the collapse of the US stock market, referred to as the 'Wall Street Crash'. This led to a loss of business confidence, as investment, output and demand fell, while unemployment rose dramatically in most western countries. A consequence of this economic hardship was social unrest, for example food riots occurred in the cities of London, Bristol and Liverpool in England. The classical economic approach to the problem was that workers would price themselves back into employment through wage cuts. However, this failed to happen, leading to a major questioning of classical economic thinking.

Macroeconomics and tourism

The inability of classical economic theory to deal with the problems of the Great Depression led to an analysis and theoretical development of the workings of the economy, which moved beyond an emphasis upon the equilibrium of individual markets, to a consideration of the equilibrium of total aggregated demand and supply. This approach, termed 'macroeconomics', subsequently considers demand, investment and supply in the whole economy rather than individual markets. As Tribe (1999: 9) comments: 'Macroeconomics looks at the economy as a whole. The national economy is composed of all the individual market activities added together.'

The origin of macroeconomics is usually associated with the work of John Maynard Keynes (1883–1946). However, as Gill (1967) points out, large parts of classical economic thinking were macro in nature, but it was Keynes that gave an enormous stimulus to this particular approach. Keynes acted as an adviser to the British Treasury and was particularly influenced by the events of the Great Depression, producing two main works, *A Treatise on Money* (1930) and *General Theory of Employment, Interest and Money* (1936), in which he challenged the assumptions of classical economics.

The basic theory of classical economics was that equilibrium within the economic system would be achieved with full employment. Thereby if demand for a product fell, labour would either shift to produce a new product, or alternatively accept wage cuts to continue working. However, based upon his observations of the Great Depression, Keynes emphasised the need to take a holistic view of the market because what happens in one market affects others. He argued that the action of cutting wages in the hope of pricing people back into work, if implemented on a large scale, serves only to drive down the general levels of incomes and consumption. This in turn reduces employment prospects further, as falling incomes and spending mean less demand, making it less likely businesses would invest in increasing their levels of production and take on more workers.

As aggregate demand in the 1930s was insufficient to keep the bulk of the labour force in employment, radically Keynes argued for expansionary government policies and interference in the markets to remedy the situation. He argued that in a cycle of economic depression, full employment could only be achieved through the stimulation of demand in the economy, and that if the private sector were unable to provide this then the government must do so. The underlying principle of Keynesian economics is that government spending should be inversely proportional to private trade (Stokes, 2002). That is, when trade is booming the government should spend little, while when the economy slumps, public investment should increase. The concept that the government should intervene in the economy to influence demand is referred to as 'demand management policy'.

The particular concerns of macroeconomic theory include economic growth; full employment; price stability and the balance of payment equilibrium (Burningham et al., 1984). It is the potential of tourism to make a positive contribution to these measures that make it attractive to many national governments. As Sinclair and Stabler (1997) point out, there can be little doubt that high levels of international tourism expenditure have a significant economic impact on tourist origin and destination countries. The typical kinds of macroeconomic benefits that governments can expect from tourism include: (i) earning foreign currency and making a positive contribution to the balance of payments; (ii) developing the service sector and contributing to the gross domestic product (GDP); (iii) attracting inward investment and income multiplier effects; and (iv) employment creation.

Balance of payments

The potential for tourism as an export industry is underlined by the fact that only exports of oil, automotive products and chemicals exceed that of tourism at a worldwide level. In 2002 tourism represented approximately 7 per cent of the total exports of goods and services, and nearly 30 per cent of service exports (WTO, 2004a). While the export potential of tourism makes it particularly attractive to less-developed countries, whose economies may be traditionally reliant on primary products, it also has an important role to play in the balance of payments in developed countries. For example, in the early 1990s the Finnish government, recognising the role tourism had to play in reducing its balance of payments deficit, commented:

> The Finnish economy is in dire need of foreign exchange revenues to narrow and eventually stabilize the yawning current account deficit. This can only be achieved through a successful economic policy and broadly based, competitive trade in exports. The tourist industry can and must do its share towards this end.
>
> (Finnish Tourist Board, 1993: 4)

Similarly, recognising tourism's potential as a foreign currency earner, the Icelandic Tourist Board (2003: 2) commented: 'Tourism is the second largest foreign currency earner after the fisheries producing 13 per cent of the country's income from foreign sources.'

Tourism plays a role to varying degrees of importance in most countries' exports, which in turn influences their balance of payments, explained by Pass *et al.* (2000: 29) as: 'a statement of a country's trade and financial transaction with the rest of the world over a particular period, usually one year.' It is essentially a measure of a country's exports of goods and services compared to its imports of goods and services and is usually divided into two main groups of 'visible' or 'invisible' items. Visibles are tangible foods, including raw materials, foodstuffs, oil and fuels, semi-processed and finished manufactured goods, so called because they can be seen and recorded by customs as they move in and out of a country (Pass *et al.*, 2000). Invisibles include earnings from and payments for services such as banking, insurance, transportation and tourism. Invisibles cannot be seen and recorded; consequently their value as exports and imports has to be determined from company returns, government accounts, foreign currency purchases and sales data from banks.

In terms of how a balance of payments specific to tourism may be explained, Youell (1998: 141) comments: 'The "tourism balance" of a particular country will consist of all receipts from its overseas visitors less payments made by its own residents on travel abroad.'

When a country has a high economic dependency upon tourism, then tourism will have an important role in its balance of payments. Governments attempt to have a positive balance of payments figure as this enables them to repay foreign loans and build up capital reserves, while a negative balance of payments is likely to lead to increased levels of borrowing and a running down of capital reserves. The balance of payments figures for tourism, for a selection of countries in 2001, are shown in Table 4.1, based upon the compilation of statistics provided by the World Tourism Organisation.

The method to calculate the balance of payments is to subtract the 'International Tourism Expenditure' from the 'International Tourism Receipts'. What actually constitutes 'international tourism receipts' are defined by the WTO (2004c: 3) as: 'the receipts earned by a destination country from inbound tourism including all tourism receipts resulting from expenditure made by visitors from abroad, on for instance lodging, food, drinks, fuel, transport in the country, entertainment, shopping, etc.' International tourism expenditure is defined as: 'the expenditure on tourism outside the country of residents by visitors (same-day visitors and tourists from a given country of origin)' (ibid.).

To explain the pattern of the balance of payments figures shown in Table 4.1 is complex, as they are a product of a range of influences. Included in these would be a country's level of economic development; levels of disposable income among members of its population; cultural attitudes to travel; natural and cultural resources for tourism;

Table 4.1 Balance of payments figures for selected countries in 2001 (sources: WTO, 2003b; WTO, 2004b).

Country	International tourism receipts (US$ million)	International tourism expenditure (US$ million)	International tourism balance (US$ million)	
			Positive	Negative
Australia	7,624	5,812	1,812	
Bahamas	1,636	297	1,339	
Barbados	687	94	593	
Belize	121	24	97	
China	17,792	17,900		−108
France	29,979	17,718	12,261	
Germany	18,422	51,900		−33,478
Japan	3,301	26,530		−23,229
Spain	32,873	5,974	26,899	
United Kingdom	16,283	36,486		−20,203
United States	71,893	60,117	11,776	

geographical proximity and travelling time to major tourism-generating countries; levels of development of tourism infrastructure; and its perceived attractiveness by people from other cultures.

Think point

Australia, France, Germany, Japan, Spain, the UK and the US are all classified as economically advanced countries. Four of the countries have positive balance of payments for tourism and three of them have negative ones. How do the factors of level of economic development; cultural attitudes to travel; natural and cultural resources for tourism; geographical proximity and traveling time to major tourism-generating countries; levels of development of tourism infrastructure; and a country's perceived attractiveness by people from other cultures, help to explain these figures?

Contribution to GDP

An important indicator of an economy's strength is its Gross Domestic Product (GDP), subsequently, one way of assessing tourism's economic significance is to measure its contribution to GDP. The GDP is a measure of the total value of all the goods and services produced in an economy over a certain time period, usually one year. Sometimes, tourism expenditure may also be expressed as a percentage of the Gross National Product (GNP). The GNP is a further refinement of the measure of GDP, including income earned by domestic residents from foreign investments, while deducting income earned by foreign investors in the country's domestic market during a measured period of time, typically three months or one year. Until fairly recently, GNP was the commonly used measure of a country's total production, but almost all countries now use GDP (Heilbroner and Thurow, 1998).

Table 4.2 The contribution of travel and tourism to Gross Domestic Product (GDP) in 2004 (source: WTTC, 2004).

Country	The contribution of travel and tourism to GDP (percentage)	Country	The contribution of travel and tourism to GDP (percentage)
Antigua and Barbuda	82.1	India	4.9
		Indonesia	10.3
Australia	12.3	Ireland	7.1
Belize	23.5	Italy	11.4
British Virgin Isles	95.2	Maldives	74.1
		Saint Lucia	47.9
China	11.4	Spain	19.9
Cyprus	27.6	Thailand	12.2
Egypt	15.3	Vanuatu	52.4
France	12.6	US	10.7
Gambia	21.6		

Tourism's contribution to the global GDP in 2003, based upon calculations that include the direct and indirect effects of tourism expenditure, is estimated to be US$ 4217.7 billion or 10.4 per cent of total GDP (WTTC, 2004). The predicted contribution of travel and tourism to the GDP of a selection of individual countries is shown in Table 4.2, based upon the direct and indirect effects of tourism expenditure, which are explained in the next section of the chapter.

As for all economic measurements treated in isolation, attention is needed not to over-generalise about the state of an economy when interpreting tourism's contribution to GDP. However, based upon the examples of Antigua and Barbuda, British Virgin Isles, Maldives, Saint Lucia and Vanuatu, it would appear that tourism is likely to be a highly significant player in the economies of 'small island developing states', contributing in these cases approximately 50 per cent or more to the total GDP. The danger of this situation is an economic over-dependence upon tourism, making the economy very vulnerable to a downturn in tourism demand. There is a danger of tourism becoming a form of 'monocrop' with few available economic options to it. However, while diversification in an economy is a sign of health, there are often limited economic alternatives for less-developed countries that may lack the technical ability to develop other resources (UNEP, 2004).

Table 4.3 Top five countries ranked in terms of international tourism receipts in 2002.

Country	International tourism receipts (US$ million)
United States	66,547
Spain	33,609
France	32,329
Italy	26,915
China	20,385

In contrast, in the cases of the US, Spain, France, Italy and China, the contribution of travel and tourism as a percentage of GDP does not exceed 20 per cent, despite these five countries being the world leaders in terms of being the beneficiaries of international tourism receipts, as is shown in Table 4.3. In contrast to smaller islands these countries possess a higher level of economic development, with a wider diversification of economic activity transcending the primary, manufacturing and service sectors, making them less dependent upon tourism.

Think point

What are the potential dangers for the economy and society of a country that is over-dependent upon tourism? What kind of changes and events can cause a decrease in tourism demand for a destination?

Tourism multipliers

To appreciate more fully how tourism impacts upon an economy it is necessary to understand a key theory of Keynesian economics, the multiplier concept. The rationale of the multiplier process is that a change in the level of demand in one section of the economy affects not only the industry that produces the final product or service, but also the other firms in other sectors of the economy that in turn supply it (Cooper *et al.*, 1998). For example, in the case of a new investment project such as a hotel, the money paid out for the goods and services needed to build it, e.g. building materials and construction workers' wages, does not stop there but continues to circulate in the economy. The recipients of this first round of investment spending will in turn enter into additional spending of their own, encouraging in turn another round of spending. So from the investment in the hotel, money could theoretically circulate a few times in the economy, generating extra demand at each stage.

Three main types of income generation are observable, the first or 'direct income' is the money invested into the economy, for example to build a hotel or from a tourist who stays in a hotel for one night. The 'indirect income', refers to the income distributed to suppliers of goods and services required by tourism enterprises (Bull, 1991). In turn, some of this money will be passed on to the suppliers of goods and services they require and so on. In addition, at each stage of circulation, money accrues to local recipients through households and firms in the form of increased incomes, which they may choose to spend or save. The money they choose to spend in the economy will generate a further round of economic activity, known as the 'induced effect'.

In addition to there being various types of income generation, owing to the various types of effects that tourism expenditure may have upon the economy, there are also different types of economic multipliers. These include the income; sales; employment; and output multipliers. Examples of the typical tourism income multiplier (TIM) are for the Republic of Ireland 1.72; Egypt 1.23; Fiji 1.07; Malta 0.68; and Western Samoa 0.66 (Cooper *et al.*, 1998). These figures refer to the total income value that is generated from the initial tourism expenditure or investment. For example, if the multiplier value is 2, then for tourist expenditure of $1,000 the total value of income created from it circulating in the economy is $2,000 (Bull, 1991). Subsequently, a figure of 1.23 for Egypt would mean for $1,000 of tourist expenditure the total value of income created would be $1,230.

Theoretically, the initial tourism investment could circulate indefinitely in the economy, but it doesn't. The reason for this is that in each round of expenditure, money leaks out of the economy, through what are termed 'economic leakages', thereby removing it from circulation in the economy. Consequently, the foreign-exchange earnings for tourism do not reveal its true economic benefit to an economy, simply revealing what remains after deducting the foreign-exchange costs of tourism (EIU, 1992). The leakage of expenditure occurs for a variety of reasons including repaying interest charges on foreign loans for tourism development; imports for the tourism industry, including materials and equipment such as kitchen equipment for hotels, and consumables e.g. food, drink and beach products; the employment of foreign workers who may send money home to their relatives; the repatriation of profits by foreign travel and tourism companies; the saving of money; and the paying of taxes to the government.

The degree of influence of these factors upon tourism income will largely be determined by a nation's or community's level of economic development. For example, many less-developed countries are reliant upon multinational corporations and large foreign businesses for the capital to invest in the construction of tourism infrastructure and facilities. Consequently, a high leakage arises when overseas investors who finance the resorts and hotels repatriate the profits to their own countries (UNEP, 2004). Larger economies also possess a more diversified structure reducing the requirement for imports to meet the needs of the tourist industry.

A worst-case scenario of the inability of an economy to meet the demands of the tourism industry is shown in Figure 4.1. Although this diagram is perhaps over-simplistic it nevertheless illustrates the potential reliance of the tourism industry upon imports.

Think point

Think about the last time you were in another country. What evidence was there of the importation of goods or other services to meet the demands of the tourists?

An additional influence upon an economy's propensity for leakages is the type of tourism market it is servicing. For example, package tours or holidays generally have a high leakage factor, with approximately 80 per cent of all travellers' expenditures going to airlines, hotels and other international companies, rather than local businesses or workers (UNEP, 2004). Local businesses will also see their chance to earn income from tourism reduced by 'all inclusive' vacation packages where the tourist remains for the entire stay in their resort or cruise ship, also known as 'enclave tourism'.

When tourists pay remittances to tour operators and airlines that are foreign to the country they are visiting, economic leakage may not be solely limited to the destination, which Smith and Jenner (1992) refer to as the 'pre-leakage' factor. A key determinant of this

Figure 4.1 Typical economic leakages (source: Seekings, 1993).

Table 4.4 Proportion of a tour operator's price that is received by a destination (source: Smith and Jenner, 1992).

Region	Percentage
South America	45–50
Egypt	35–50
North India	35
China	30–35
South India	20

pre-leakage factor will be the use or non-use of the national airline carrier. The cost of air transport is usually a substantial percentage of the total cost of a travel vacation package, typically in the region of 40 to 50 per cent. Subsequently, if the money for this service is paid to an airline foreign to the destination the pre-leakage factor will be high. Typical pre-leakage factors for different regions of the world are shown in Table 4.4.

From the figures given in Table 4.4, it is evident that the majority of the tourist's expenditure never arrives in the destination. For any government hoping to maximise the economic benefits of tourism, reducing the causes of leakage would be a central aim. In an attempt to reduce the pre-leakage factor, encouragement of the use of their national carrier by the tour industry and tourists is important. However, this may be especially problematic for smaller developing countries and islands, where the economics of the airline business and the control of routes by established airlines dictates against the establishment of a national carrier.

Governments may also consider the imposition of taxes or duties on imported goods for tourism, e.g. food and beverages, where there is an alternative product substitute available in the local economy. This would, in theory, make local goods more price competitive and stimulate local economic demand. However, such a strategy is dependent upon having a production capability in the economy able to meet the quantity and quality demands of the tourism industry. Other potential methods might include restrictions on foreign investment and on the employment of non-nationals (EIU, 1992).

However, controls on the repatriation of profits by multinationals; measures to ensure foreign investment is only allowed in partnership with local companies in an attempt to make sure that a higher percentage of profits remain in the country; and the encouragement of local industry through the use of import duties and tariffs are increasingly difficult in the contemporary era of 'free trade' that encourages the free flow of capital between countries. An alternative strategy to government measures to control leakages could be to develop types of tourism that have a lower leakage factor. For example, the encouragement of the kinds of tourism where tourists want to sample local goods and services, rather than imported ones, offers an opportunity to reduce the leakage factors. Typically, this kind of scenario is equated with the development of alternative types of tourism such as ecotourism, nature tourism, and activity-based hotels. However, the use of such a strategy does not necessarily overcome the problems of the pre-leakage of tourism expenditure, and would also be likely to lead to a reduction in the overall level of tourism earnings in countries with already established tourism industries.

Think point

How could the types of leakages shown in Figure 4.1 and pre-leakage factors be reduced?

Tourism employment

A major economic attraction of tourism for governments is its potential for employment creation. As Seekings (1993: 13) comments:

Often faced with high levels of unemployment or underemployment, most governments are concerned to establish activities that will provide regular, productive jobs. Tourism is a labour intensive service industry which creates a high level of job opportunities of a type suitable for a semi-skilled and unskilled labour force.

An added attraction of employment creation through tourism is that it is held to require less investment than in the manufacturing sector (Youell, 1998).

Tourism is a major contributor to global employment, with an estimated 214,697,000 or 8.1 per cent of total global employment in 2004, directly and indirectly attributable to its economic impact (WTTC, 2004). Similar to incomes, the multiplier effect also works for employment, generating extra jobs not only in the tourism sector, but also indirectly in other sectors of the economy. The percentage of a country's workforce employed directly and indirectly in tourism is indicative of tourism's importance to that country's economy.

Statistics for different countries showing the employment generated directly and indirectly by tourism as a percentage of the total workforce in 2004 are shown in Table 4.5.

Think point

Analysing Table 4.5, for which five countries is tourism the most significant contributor to employment? For which five is it the least significant? Attempt to explain this pattern.

Table 4.5 Estimated employment in tourism in 2004 (source: WTTC, 2004).

Country	Percentage of total employment generated directly by travel and tourism	Percentage of total employment generated directly and indirectly by travel and tourism	Country	Percentage of total employment generated directly by travel and tourism	Percentage of total employment generated directly and indirectly by travel and tourism
Antigua and Barbuda	35	95	Jamaica	11	31.8
Barbados	20	58.3	Japan	4	10.3
Belize	9	23.1	Macau	33	79.1
British Virgin Isles	38	95	Malta	20	34.7
Croatia	14	28.9	Slovenia	5	16.7
Cyprus	19	35.9	Sweden	2	7.2
Estonia	5	20.7	United States	5	11.9
France	6	15.1			

Similar to GDP, there is the danger of creating an over-dependency on tourism as a source of employment. This is particularly the case for small islands where the opportunities for the development of other economic sectors may be limited by resources. However, an over-dependency may also be encouraged by the comparatively high financial rewards and salaries from tourism in comparison to other sectors of the economy, which may result in a shift in workers from other economic sectors to tourism. For example, in 1981 on the Greek island of Corfu, 41 per cent of the workforce was employed in the primary sector, mainly in agriculture, and just 10 per cent in tourism. By 1990 the percentage of the workforce employed in agriculture had reduced to 12 per cent, while in the tourism sector it had risen to 24 per cent (Tsartas *et al.*, 1995).

Similar to other sectors of the economy, productivity gains through the use of technology may negate the reliance upon firms to employ more workers. Also, as a tourism sector develops, there may be an increased reliance upon imported goods, thereby reducing the opportunities for employment generation (Burns, 1993). The skills levels of the workforce and opportunities for tourism education and training will be critical in determining the quantity and types of opportunities for local people to work in the industry. The establishment of hotel and tourism training schools is necessary to avoid an over-reliance on foreign staff, and to give local people the knowledge and skills necessary for opportunities to work in the industry. A further consequence of a shift in the balance of employment is the loss over time of the knowledge and skills necessary to keep other parts of the economy functioning.

Criticisms of the quality of employment in tourism have also been made, including that it is often low-skilled; has a lack of union recognition; is part-time and seasonal; and also in developed economies has poor levels of pay compared to other types of work. Cultural conflicts may also be caused, notably in less-developed countries where the best employment opportunities are given to expatriots rather than local people. This has been particularly noticeable in the Caribbean and parts of Africa, where many hotel management jobs have been given to expatriot white staff rather than to local people, raising the issue of racism.

Further cultural conflict may also arise when tourism employment is felt to be servile, especially in cultures where there is a tradition of slavery, or a history of colonisation by western countries. Additionally, the nature of tourism as a service-based activity means that employees may be required to work irregular hours, which may lead to conflicts over the observation of religious practices, e.g. going to church, the mosque or gudwara, or over expected domestic household duties and roles. This is likely to be more noticeable in patriarchal societies if tourism provides employment opportunities for women. If it has not been the tradition for women to work, the financial and social empowerment of women through working in tourism may lead to a challenging of their traditional roles and duties, a theme that is developed more fully in the final chapter.

Problems of measurement

A key issue in attempting to make sense of the economic impact of tourism is the accuracy of available economic data. For example, concerning the impact of tourism upon employment, the WTO (1998: 87) commented:

The contribution of tourism to employment and tourism's potential to generate new jobs ranks as one of the paramount questions related to the social and economic importance of tourism. However it is difficult to make accurate assessments of its volume and impact on the economy. Unfortunately reliable and comparable data about tourism employment on the international level is very scarce.

An accurate economic assessment of the impact of tourism is important because it forms the basis of the decision-making of government on whether to allocate resources to tourism. One infamous example of where a faulty economic assessment was made over the economic impact of tourism was the 'Zinder Report' (Lea, 1988). Zinder and Associates, who were employed by the United States Agency for International Development, calculated the TIM for the Eastern Caribbean Isles to be 2.3, a calculation that gained acceptance by many governments. However, when subjected to a detailed critique by other economists, the calculation was found to be a large overestimate of the income multiplier. The reason for this error was the summing of successive 'gross receipts' at each round of spending rather than using the successive 'incomes received' by residents (Lea, ibid.). Calculating the TIM this conventional way produced a multiplier figure of only 1.36 instead of 2.3.

Difficulties in the collection of comparable data are partly a consequence of the nature of tourism, in the sense of it being disparate, and being composed of different industries, e.g. airline industry, hotel industry, attractions industry, tour operating industry. The disparate nature of the tourism industry means that tourism expenditure is subsequently spread across several sections of the economy, necessitating accurate surveys of visitor expenditure to break it down into its various components, e.g. accommodation, meals, beverages, transportation and shopping (Cooper *et al.*, 1998). Different countries of the world using different data collection methodologies to attempt to measure the same characteristics compound this problem. In addition, the fact that many tourism businesses are small and run by the self-employed means they may not appear in readily available tourism statistics (Youell, 1998). These difficulties are pointed out by the WTTC (2004: 6) who comment: 'Over the last three decades, countries have estimated the economic impact of Travel and Tourism through a range of measures using a variety of different definitions and methodologies.'

This lack of uniformity in collecting data about the economic impacts of tourism has made it difficult to make a meaningful comparison of tourism's role in the economies of different countries. As a consequence of this lack of coherence, and a commonality in economic definition and methods for data collection, the WTO have been working together with the WTTC since the early 1990s to improve tourism statistics worldwide (WTTC, 2004). The common framework that is being developed is termed the 'Travel and Tourism Satellite Account', which lays out a comprehensive recommended methodological framework for the compilation of tourism statistics and the assessment of its economic importance in all countries. The adaptation of this system by different countries, for example the information in Tables 4.2 and 4.3 is based upon it, should permit a more meaningful comparison of tourism's role in different national economies.

Liberalisation, tourism and trade

While trade in international tourism has become established as a major part of the global economy, how this trade is carried out is subject to influences that are often beyond the control of an individual country, but which will determine the economic impacts tourism will have in it. The most influential development in international trade during the last two decades has been the trend to 'trade liberalisation' and 'free trade'. Liberalisation includes the removal of any measures that a national government may wish to pursue to protect its own industries, such as tariffs, quotas or foreign-exchange controls, and the deregulation of its economy, removing any restrictions on the foreign establishment, ownership, employment of personnel, and remittances. The aim of liberalisation and deregulation is the creation of a 'free market' that operates at a global level. In the case of tourism, this will mean that foreign owned and run companies will have freer access to domestic markets under the same trading conditions that exist for local companies of the host country.

A chief advocate of the free-trade system is the neo-classical economist Milton Friedman, who greatly influenced the political thinking of Ronald Reagan and Margaret Thatcher in the 1980s, who in turn were the major political advocates of the free-market system. International trade in tourism is regulated through the General Agreement on Trade in Services (GATS). This agreement encourages countries to liberalise their economies, making them open to foreign investment and the free flow of capital. While this is likely to provide new investment opportunities for large multinationals, the benefits for less-developed countries are less certain. However, many less-developed countries in the world are forced to accept the liberalisation of their economies, as part of a 'structural adjustment programme'. It is designed to improve a country's foreign investment climate by eliminating trade and investment regulations; boosting foreign exchange earnings by promoting exports; and reducing government deficits through spending cuts. Typical measures include: currency devaluation; trade liberalisation; cuts in government spending on health, welfare and education; the privatisation of government-owned enterprises; wage suppression; and business deregulation.

The aim of free trade is to secure the benefits of international specialisation (Pass *et al.*, 2000). The basis of this idea is that each country should specialise in a certain number of economic activities for which it possesses the best natural and human resources. In theory, specialisation enables an economy to make the most efficient use of its scarce resources, thereby producing and consuming a larger value of goods and services than would otherwise be the case (Pass *et al.*, 2000).

Specialisation is associated with the theory of 'comparative advantage', developed by the renowned classical economist David Ricardo (1772–1823). The basis of this theory is that every country in the world has an economic advantage of some kind or another in which they should specialise. Through trading, each country could specialise in what they are best able to do and maximise their resources, rather than trying to produce everything they need. Using the analogy of the family, Mankils (2001) points out that although families are competitive with each other, e.g. for employment or shopping, it is unlikely a family would be better off by isolating itself from other families and markets. If it did, they would have to grow their own food, build their own shelter

catalyst for development will be determined not only by economic theories, but also by the existing power structure between and within countries. This is especially important, as tourism has become viewed by international agencies and many national governments as a means for development. However, although the term 'development' is a common one, what does it mean and how did it become synonymous with economics?

What is development?

Why should tourism be talked of as a form of development? What is meant by development? How does the concept of 'development' manifest itself at the global and national scales? These are critical questions to understanding tourism's place in the global system. 'Development' is a term that is often used in a variety of different contexts. For example, a common usage is to talk of an individual's development or perhaps lack of it. Often this implies some change in their behaviour, intellectual capabilities and skills, or physique. The theory of human development expressed in Darwin's *Origin of the Species* in the nineteenth century coincided with the great economic and social changes of the Industrial Revolution also referred to as development. Philanthropists and liberals, as described in Chapter 1, shared concerns over the appalling living and working conditions of many people in this period. This led to observations that society may also pass through evolutionary stages of development similar to the human body, until a stage of utopia is reached.

However, it was not until after the end of the Second World War in the 1940s that a strong association was established between economic progress, the development of society, and politics. This period was marked by the ideological divide between the political and economic model of capitalism with the alternative model of communism. It was also a time when countries were achieving political independence in the post-colonial period in an era of marked change in the world political map and balance of power. A major post-Second World War concern among the leaders of the 'allied nations', including the US and UK, was to establish a system of rules and regulations to govern the post-war global economy. A particular concern was to avoid the economic hardships of the 1930s 'Great Depression'. To help to achieve this aim a meeting of international leaders was convened in Bretton Woods in New England in 1944, which resulted in the establishment of institutions to regulate the world economy, the details of which are discussed in detail in Box 5.1.

A further significant event in terms of establishing the terminology of international development was President Truman's speech on his inauguration as President of the United States of America in 1949. In this speech he distinguished between the 'developed' and the 'underdeveloped' worlds, and the 'First' and the 'Third' worlds, terms defined by the quantitative measure of national income and political orientation. The First World encompassed the industrialised and capitalist countries of the world, typically those of North America, Western Europe, and Australasia. The 'Third World' countries were the economically developing countries of the world and politically non-aligned, typically those of the continents of Africa, Asia and Latin America. In between the First World and the Third Worlds was a 'Second World' that consisted of the industrialised coun-

Box 5.1 Bretton Woods

In 1944, delegates from 44 nations met at the New England Resort of Bretton Woods, to devise a new framework for the global economy. They were concerned to try and stabilise the international economy to avoid a repeat of the Great Depression that had proved economically and politically damaging. An important member of the delegates was the economist Keynes who was referred to in Chapter 4 as a key influence on the development of macroeconomics. The title 'Bretton Woods Trio' refers to the three governing institutions which emerged from the conference to coordinate the global economy. These are the IMF, the World Bank and GATT.

The International Monetary Fund (IMF)

The role of the IMF was to create economic stability in the world and to oversee a system of fixed exchange rates. The idea of this was to stop countries devaluing their national currencies to gain a competitive trading advantage over other countries, a strategy that was believed to have been a major factor contributing to the global economic downturn in the 1930s. The IMF was to provide limited loan amounts to members when they fell into a trade deficit.

The World Bank (International Bank for Reconstruction and Development (IBRD))

Another key goal of the Bretton Woods conference was to rebuild the economies of Europe after the Second World War. The IBRD was established to facilitate this process, being funded from payments from its members and money borrowed on international money markets. Emphasis was placed upon the giving of loans for the development of infrastructure, including dams, roads, airports, ports, power stations, agricultural development and education systems. However, although the bank lent large amounts of money for the redevelopment of Europe, the reconstruction of Europe was also facilitated by the US through the form of grants in what was termed the 'Marshall Plan'.

As Europe's economy gradually recovered in the 1950s, the attention of the IBRD moved towards the 'Third World' and it became widely known as the World Bank. Linked to the 'stages of growth' economic theory, a strong infrastructure was essential for economic development. Consequently, the Bank funded hydroelectric projects and highway systems throughout Latin America, Asia and Africa. However, it was soon clear that the poorest countries would find it difficult to meet loan repayments. In the late 1950s the Bank established the International Development Association (IDA) to provide 'soft loans' with very low rates or no rate of interest at all.

General Agreement on Tariffs and Trade (GATT)/World Trade Organisation (WTO)

The aim of GATT was to establish rules to govern world trade and reduce national barriers to trade. Seven rounds of tariff reductions were negotiated under the GATT treaty, the final Uruguay Round being in 1986. In 1994, a new organisation – the World Trade Organisation – was established in Morocco, which had the official status of an international organisation rather than a loosely structured treaty. It includes the GATT agreements, and also a new General Agreement on Trade in Services (GATS), which includes banking, insurance, telecommunications and tourism, as explained in Chapter 4. Both GATT and the WTO have been subsequently criticised as a 'rich man's club' dominated by the western nations on the basis that the global trade rules are biased against developing countries.

Source: Ellwood (2001)

tries of the Soviet Union and other communist allies. However, since the collapse of the Soviet Union post-1989, these terms have little validity applied to the contemporary global political and economic situation, and the use of the term 'Third World' is now viewed as being derogatory, developing countries being preferable.

Truman's speech was the watershed in terms of defining development as an economic measurement. Subsequently, as 'development' became associated with economic progress, it ignored other types of wealth, as Reid (1995: 140) comments: 'The wealth of a range of societies with flourishing cultures was simply ignored . . . for it could not be assessed in economic terms.' Consequently, the dominant measure of development was material wealth, ignoring the range of democratic, gender, freedom and environmental issues that also accompany economic changes.

Think point

What do you think 'development' is? How would you attempt to explain what 'wealth' is?

In the context of this economic definition of development, logic would also suggest if a country is underdeveloped, it requires a growth in its economy. However, other reasons may exist to explain the state of comparative underdevelopment. Notably the power relationships that exist between nation states and different groups within society, in political and economic terms, influence how resources are distributed. This is a central theme of 'dependency theory', explained in the next section of this chapter.

The establishment by President Truman of an American interest in underdeveloped countries of the world as part of US policy also invited comment about its political and economic intentions. Some commentators have noted that the desire to 'develop' countries of the world was associated with the expansion of the capitalist model into new markets, at a time when the world was dividing on ideological lines between the capitalist and communist systems.

A theoretical framework for development

The changing political and economic climate of the post-Second World War period marked the beginnings of the study and theorising of development, alongside the formalisation of policy at national and international levels. Different frameworks of main ideas or 'paradigms' of development have been formulated in the post-war period. These include modernisation; dependency; economic neoliberalisation; and alternative/ sustainable development theories, as are shown in Figure 5.1. It is necessary to emphasise that the timeframes act only as guidelines to when a paradigm gained prominence

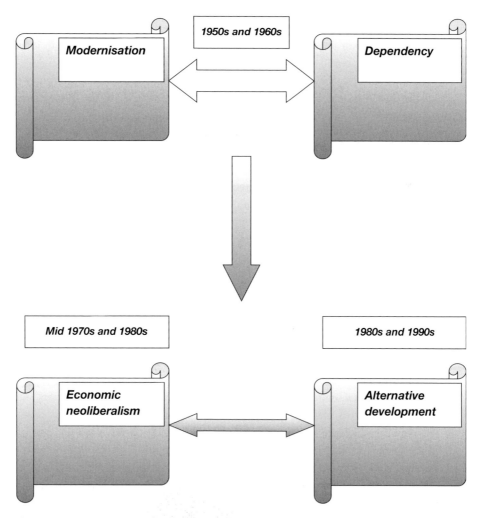

Figure 5.1 Development theory and paradigms.

after the Second World War, with many components of the different theories influential in contemporary policy formulation.

Modernisation

The association between the notion of organic development and the development of societies manifested itself in the first major theoretical perspective on international development, that of modernisation theory. Modernisation views socio-economic development as an evolutionary and linear path from a traditional society to a modern society. Its roots lie in a variety of different perspectives applied by non-Marxists to developing countries in the 1950s and 1960s, but it is epitomised by the work of Walt Rostow, who was a policy adviser to President Johnson of the US. The title of his most influential work, *Stages of Economic Growth – A Non-Communist Manifesto* published in 1960, indicates the political context of Rostow's work. According to Slattery (1991), Rostow's work extended beyond the theoretical into a political manifesto to fight off the threat of communism, or the 'disease of transition' as he referred to it. Consequently, in Rostow's view, financial aid for development should surpass the purely economic to include the political, i.e. supporting non-communist elites, democracy and pluralism.

The work of Rostow was heavily influenced by the organic analogy of functionalism, i.e. the idea that societies, like natural beings, mature through different stages of evolution driven by an internal dynamic. The different stages of Rostow's model are shown in Figure 5.2.

Rostow outlined in detail the five main stages of economic growth that, in his view, societies must pass through; in his belief this was inevitable, with the momentum being maintained by the dual effects of an increasing population and the rise of modern living standards. In his modernisation model, Rostow suggested that rapid economic development could occur only if barriers to tradition and superstition could be overcome, and the values and social structures of traditional societies are changed (Harrison, 1992). As Slattery (1991: 270) comments:

> Just as we often talk of a child being immature, even retarded, when its mental development doesn't match its physical growth, so we talk of underdeveloped even backward societies, of Third World countries held back by illiteracy, ignorance and superstition.

In Rostow's view such a change would involve the development of features such as investment capital, entrepreneurial skills and technical knowledge. He further suggested that if such features are absent, possibly because of the conservative nature of tradition, they could be diffused from outside, or even historically through colonialism. Constant to the theme of modernisation is one of 'westernisation' in which the structures of less developed countries become like those of the West, emulating their development patterns (Harrison, 1992). In this sense it can be argued that Rostow appears to justify colonialism (Harrison, 1988), on the basis that the European powers often included modernisation as one object of colonial policy.

STAGE ONE – TRADITIONAL SOCIETY
Pre-industrial, usually agricultural societies.
Characterised by restricted and low output, ancient
technology and poor communications. A
hierarchical social structure with little social
mobility. Values are fatalistic.

STAGE TWO – PRECONDITIONS FOR TAKE OFF
New ideas favouring economic progress arise, leading
to the idea of economic change through increased
trade and the establishment of infant industries. A new
political elite emerges, such as the entrepreneurial
bourgeoisie of the Industrial Revolution to challenge
the power of the landed classes.

STAGE THREE – THE TAKE OFF
Industrialisation replaces agriculture as the
generating force of the economy. This
happens through new technology or the
influence of the entrepreneurial class who
prioritise modernisation of the economy.
Agriculture also becomes commercialised
with a growth in productivity. The influence
of the market leads to new political, social
and economic structures. Rural to urban
population drift takes place.

STAGE FOUR – THE DRIVE TO MATURITY
Over a period of approximately 40 years, the
country builds on its progress. Investment
grows, 10–20 per cent of the national income
is invested in industry, technology spreads to
all parts of the economy, and the economy
becomes a part of the international system.
There is a move away from heavy industry, as
what is produced now becomes a matter of
choice rather than necessity.

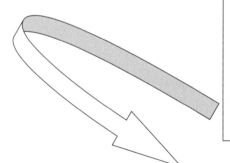

STAGE FIVE – THE AGE OF MASS CONSUMPTION
The economy matures, the population can enjoy the benefits of mass consumption, a high
standard of material living and, if it wants, a welfare state. According to Rostow, in the US this
stage was symbolised by the mass production of the motor car. The balance in the
economy progressively shifts to a service economy vis-à-vis an industrial one.

Figure 5.2 Rostow's 'Five Main Stages of Economic Growth'.

Given its political emphasis and western perspective, unsurprisingly, modernisation theory has received substantial censure. These include the criticism that the uni-directional path of development suggested in Rostow's model is incorrect. The later-developing countries can learn from the mistakes of the West, and through borrowing western skills, technology and expertise subsequently leap a stage or two. Certainly in the case of tourism, any lesser-developed country proposing to use tourism as a means for development would by-pass the heavy industrialised process outlined by Rostow. This in turn is a reflection of the importance of tourism in international trade, and that the countries which generate the majority of international tourists have them-selves passed into the 'age of mass consumption' in which tourism is a popular purchase. However, at the time when Rostow was writing, the idea of considering tourism as a form of international trade and being used as a means to modernise economies would probably have seemed improbable. The path of development has been criticised for its assumption that traditional values are not compatible with modernity, and that is does not consider alternative or traditional models of development. The model has been further criticised upon environmental grounds, that if the developing world begins to industrialise in the same way as in the West, the strain on nature's resources may be insupportable.

Think point

Explain modernisation theory in your own words. Summarise its main points.

Dependency theory

The criticisms of modernisation theory were encapsulated in an alternative school of thought, known as 'dependency theory'. In a radically alternative view to modernisa-tion theory, dependency theorists argue that developing countries have external and internal political, institutional and economic structures that keep them in a dependent position relative to developed countries (Bianchi, 1999). Subsequently, development theorists attempted to formulate an explanation of the causes of underdevelopment in an holistic framework, based upon the interaction of economic and social structures within an international system.

Based upon a neo-Marxist perspective of the international political and economic system, development theory is also referred to as 'world systems theory' and 'under-development theory' (Bianchi, 2002). Emphasis is placed upon analysing the exploita-tive nature of the exchange relations between historically powerful metropolitan states, such as the US and many Western European countries, and their dependent satellites, for example countries of Latin America, Africa and parts of Asia. A major influence in this way of thinking about development was Andre Gunder Frank, who viewed devel-opment and underdevelopment as part of the same world capitalist system. Frank argued that the lack of progress in developing countries was because the western nations deliberately underdeveloped them, not because of their own inadequacies or a failure to develop a 'culture of enterprise'. The typical characteristics of the underdevelopment Frank was referring to are shown in Box 5.2.

Frank emphasised that this relationship of exploitation and dependency can be traced back to the seventeenth century and colonialism, when European powers conquered and colonised the continents of Africa, Asia and Latin America, making them a part of their imperial system. Subsequently, with the Industrial Revolution and the emergence of a world capitalist system, the development of an international division of labour forced colonies to specialise in the production of one or two primary products. The colonies supplied the mother country with cheap raw materials and food and in turn provided markets for manufactured goods. For example, an essential component of the beginnings of the Industrial Revolution in Britain was the importation of cotton, which was then made into clothes for sale in the Indian market.

Frank argued that through this model emerged a 'world system' of dependency and underdevelopment in which core nations, such as Britain, exploited 'peripheral ones', such as India. A key element within this relationship was the city, as colonial powers built cities, or used existing ones, as a means of governing areas and forging points of connection with ruling local elites. The ruling elites generally collaborated with colonial powers, using their control of local markets to exploit the peasantry in the countryside, progressively becoming more linked to a western way of life than the majority of their own people.

Frank (1967) also suggested that to preserve their rule and lifestyles the military might be used against local people at times of political unrest, with arms supplied from the West. There existed therefore a system of 'metropolises and satellites', in which development in one part of the world system occurred only at the expense of another part. The 'centres' or 'metropoles' exploit 'peripheries' or 'satellites', through the mechanism of unequal exchange, thus transferring value from the relatively underdeveloped to the relatively developed regions (Harrison, 1992). Subsequently, instead of aiding their development, the world system actually hinders the development of less-developed

---

Box 5.2 Characteristics of underdevelopment

- In the twentieth century world population increased from one billion to six billion.
- It is predicted that the world population will be 8.5 billion by 2025.
- There are 1,300 million living on less than one US$ per day.
- A further 1,600 million live on less than two US$ per day.
- Forty per cent of all children in the developing world under the age of five are underweight or starving.
- World consumption rose from US$1.5 trillion in 1900 to US$24 trillion in 1998, but the proportion of GNP per capita of low-income countries as a percentage of that in high income countries has reduced from 4 per cent in 1950 to 1.4 per cent in 1997.

Source: *New Internationalist* (1999)

countries. Since the decline of military colonialism, it is arguably multinational organisations that govern this system through economic colonialism, with profits going the West. Especially important in the continuance of this unequal relationship is the interaction of multinationals and governments with elite counterparts in the less-developed countries, and the maintenance of special trade relations (Lea, 1988).

Like modernisation theory, dependency theory has also received substantial criticism from both the political 'right' and the political 'left'. Criticisms include that its theoretical concepts are vague and its ideas are too radical and too Marxist. Modernisation theorists argue that multinationals and western aid bring considerable benefits, and that the First World needs the developing world to grow and industrialise as a source of new investment and new markets, rather than for it to remain in an underdeveloped state. The historical analysis of Frank's work is also criticised as being too generalised and failing to recognise the differences between countries and levels of development. Marxists have also criticised Frank's failure to provide a revolutionary programme for how countries can break free of the world capitalist system.

Think point

Explain development theory in your own words. Summarise its main points. How would development theorists be likely to explain the pattern of international tourism?

Economic neoliberalism

A major influence upon the international system of trade, although, as Telfer (2002) points out, a paradigm of development that has received less explicit attention than the other three, is 'economic neoliberalism'. The ideas of economic neoliberalism arose as a reaction to the threats of restricting supplies of oil from the Middle East to the West, the international debt crises in the 1970s and 1980s, and because of a lack of confidence in government planning as a means of solving underdevelopment and poverty. Emphasis was placed upon privatisation and free markets, the World Bank being a major proponent of this perspective along with leaders of the US and Britain. As Mowforth and Munt (2003: 34) comment:

> The rise and hegemony of neoliberalism symbolised by the Reagan–Thatcher axis, captured by Reagan's 'magic of the market' speech at the 1981 North–South Conference in Mexico, was characterised by the dominance and application of free market principles and trickle-down growth.

The concept of 'trickle-down growth' is epitomised by heavy capital investment in major construction projects such as dams, bridges, roads or large tourism complexes. The economic rationale for this policy is that different multiplier effects should work through the economy, generating extra income, jobs and sales. Thus the economic benefits should 'trickle down' to all classes and segments of society from the initial investment.

However, it is also the case that 'trickle-down effect' can be viewed as an integral part of modernisation theory.

Brohman (1996) comments that neoliberals take the view that developing countries, during their initial stages of development, should specialise in primary exports to emphasise their 'comparative advantage', rather than trying to attempt to develop more sophisticated industrial sectors with the aid of state resources which would not provide comparative advantages. Such outward-orientated neo-liberal development strategies have been supported by the International Monetary Fund (IMF) and World Bank, through structural adjustment policies as explained in Chapter 4. Thus, access to loans and sources of finance provided by the IMF is conditional on the adoption of policy reforms designed to reduce state economic intervention and generate market-orientated growth. This reduction in state economic intervention includes the removal of any kind of protective measures to fledgling industries, including tourism.

Brohman (ibid.) says that in many countries this emphasis on the production of primary products for export markets has led to a shift in development strategy away from an inward perspective towards an outward orientation. This includes the expansion of previously ignored sectors such as international tourism. International tourism is being grouped with other new 'growth' sectors, e.g. non-traditional agricultural exports to western countries, which are believed to show much promise for stimulating rapid growth based on the comparative advantages of developing countries.

The role of tourism as an export industry and as a means of earning foreign exchange is strongly supported by multinationals as they continue to attempt to secure new markets for their products. They also wish to have unimpeded access to resources (Scheyvens, 2002), which includes natural, cultural and human ones. Some developing countries have also wanted to increase tourism as a consequence of falling world commodity prices during the 1980s and 1990s, and the requirement to fulfil debt repayments to the IMF and World Bank.

Think point

How is tourism being used as part of economic neoliberalisation? What are the potential dangers of governments having to give foreign investors unimpeded access to a nations natural and cultural resources?

Alternative and sustainable development

The type of development supported by the Bretton Woods Trio and western governments, encapsulated within the modernisation and neo-liberal theories, has traditionally been based upon economic growth and top-down diffusion. In contrast to these paradigms other strategies of development have been advanced, collectively called 'alternative development'. As Telfer (2002: 47) comments: 'The alternative development paradigm is a pragmatic, broadly based approach, which arose out of criticism of these models.'

More specifically, the alternative paradigm is centred upon people and the natural environment, emphasising democracy and planning from the 'bottom-up' rather than the 'top-down'. This emphasis has resulted in new methods for assessing development that extend beyond simple economic measurement, in a recognition that the quality of life includes other variables than purely the economic. Examples of the more recent methods for assessing development include the Human Development Index and the Human Poverty Index compiled as part of the United Nations Development Programme. As Mowforth and Munt (2003) emphasise, these new systems of social, environmental and economic quantification represent a move towards peoples-focused and participatory approaches to development.

One alternative approach to development is to aim to fulfil basic human needs and place emphasis on the development of the human personality. Consequently, direct attacks are made on problems such as infant mortality, malnutrition, disease, literacy and sanitation. In the planning process, emphasis is placed upon indigenous theories of development as they incorporate local conditions and knowledge systems, rather than purely western models of development.

The alternative paradigm also places a strong emphasis upon the conservation of natural resources and ecosystems. The linking of human development with the conservation of natural resources is epitomised in the concept of 'sustainable development'. This is a central theme of alternative development, as within the context of tourism is 'sustainable tourism'. The term 'sustainable development' is usually associated with the Brundtland Report, officially the report of the World Commission on Environment and Development (WCED, 1987).

The Brundtland report was based upon an inquiry into the state of the earth's environment led by Gro Harlem Brundtland, the Norwegian Prime Minister, at the request of the General Assembly of the United Nations. Concern over the effects of the pace of economic growth on the environment since the 1950s led the United Nations, in 1984, to commission an independent group of 22 people from various member states, representing both the developing and developed world, to identify long-term environmental strategies for the international community (Elliott, 1994).

Accompanying a heightened awareness of environmental problems was also a realisation that the environment and development are inexorably linked. Development cannot take place upon a deteriorating environmental resource base; neither can the environment be protected, when development excludes the costs of its destruction. The way development has been pursued, characterised by a general lack of concern for the environment, has led to the use of natural resources in a way that is unsustainable, i.e. many finite resources are being exhausted while the capacity of the natural environment to assimilate waste is being exceeded.

It was the predominance of the negative aspects of these changes that led to the calls for sustainable development as an integral part of an international policy for development. The term gained greater attention following the United Nations Conference on Environment and Development (UNCED), held in Rio de Janeiro in June 1992, popularly referred to as the 'Earth Summit'. At the Earth Summit a programme for promoting sustainable development throughout the world, known as Agenda 21, was adopted by participating countries. Agenda 21 is an action plan for the twenty-first century laying out the basic principles required to progress towards sustainability. It envisages national sustainable development strategies, involving local communities and people in a 'bottom-up' approach to development, rather than the 'top-down' approach which has typically characterised national development plans.

In the last decade of the twentieth century, the term 'sustainable development' became widely used by governments, international lending agencies, non-governmental organisations (NGOs), the private sector and academia. The fact that the term can be readily adopted by such a diverse range of organisations, some of whom could be viewed as having divergent and politically opposed objectives, is a reflection of the inherent ambiguity of the concept. This ambiguity permits a variety of perspectives to be taken on sustainability. Much of this ambiguity can be traced to the most commonly quoted definition of sustainable development taken from the Brundtland Report:

> Yet in the end, sustainable development is not a fixed state of harmony, but rather a process of change in which the exploitation of resources, the direction of the investments, the orientation of technological development, and institutional change are made consistent with future as well as present needs.
>
> (WCED, 1987: 9)

Richardson (1997) describes this definition as political fudge, aimed at compromising the opposing views of commissioners from different states to keep everyone happy. However, the remainder of the Brundtland Report emphasises that, within the scope of this definition, other key issues relating to development have to be addressed, such as the alleviation of poverty, degradation of the environment, and issues of intra- and inter-generational equity.

A central theme of the Brundtland Report is that poverty alleviation through sustainable development is critical for the long-term environmental well-being of the planet. Poverty is a major cause of environmental destruction, a relationship that is particularly exasperated in regions of the world where the population is growing rapidly, and forced into more marginal environments. This link is emphasised by Elliott (1994: 1) in the following passage:

> In the developing world, conditions such as rising poverty and mounting debt form the context in which individuals struggle to meet their basic needs for survival and nations wrestle to provide for their population. The outcome is often the destruction of the very resources with which such needs will have to be met in the future.

A major issue of sustainable development rests on what are the ultimate goals and, if these can be agreed among stakeholders, how can they be achieved? This leads to a range of political opinions, including those who anticipate it can be achieved within the paradigm of neoliberalism, to a more radical perspective demanding a complete restructuring of society. This would include addressing the root causes of non-sustainability, including the distribution of power and wealth, the roles of transnational corporations, class-based politics, and gender inequalities. Thus, radical approaches to sustainable development challenge the values and principles of capitalist society, as Doyle and McEachern (1998: 37) comment: 'Radical environmental political theorists are involved in paradigm struggles, each seeking to create new sets of key values and principles that directly challenge existing, powerful paradigms.'

Political tension over sustainable development therefore underlies much of the debate about its interpretation. A fundamental division is between those for whom sustainability represents little more than improving technology and environmental accounting systems, while preserving the status quo of existing hierarchies and power structures in society, and those who have more radical political agendas involving changing the value systems and power structures of society.

Think point

Summarise the main points of sustainable development. How are the principles of sustainable development different from those of economic neoliberalism?

Development and tourism

While the field of development theory has evolved strongly during the last five decades, and tourism's potential to contribute to development has become increasingly evident, the linkage of development theory with tourism development is limited (Bianchi, 2002; Hall, 1994; Telfer, 2002; Wall, 1997). Bianchi (1999) suggests that the lack of attempts within the tourism community to engage with the paradigmatic debate of development studies is due to an emphasis on studies of an applied and practical nature. In an attempt to give a possible framework to the relationship of development theory to tourism, a suggested model is shown in Figure 5.3, based upon the work of Telfer (2002).

The first official United Nations Development Decade – the 1960s – was characterised by optimism, the general assumption being that the development problems of the less-developed world would be solved quickly through the transfer of finance, technology and experience from developed countries to less-developed ones (Elliott, 1994; World Bank, 2000). Tourism formed part of this optimism for economic development as Srisang (1991: 2) comments:

> In the late 1950s, the World Bank enthusiastically prescribed tourism development as a top economic policy for Third World governments. UN specialist agencies such as the United Nations Development Programme (UNDP), the International Labour Organisation and governments of major industrialised countries decided to back the World Bank's plan. Before long a new UN related agency, the World Tourism Organisation (WTO), was created to promote tourism development.

Consequently, during the 1960s tourism was equated with development as part of the modernisation paradigm. Wall (1997) equates this era with the 'trickle-down effect',

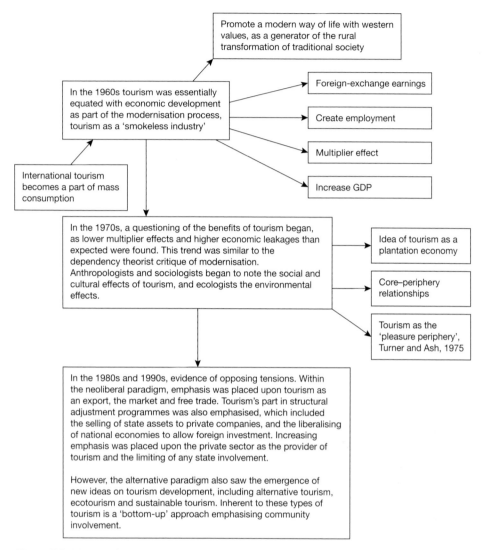

Figure 5.3 History of tourism development.

that is, the idea that investment in large-scale development will lead to economic benefits being passed down to the lower social classes of society, in a 'top-down' approach
to development. There was a belief that tourism could earn foreign exchange and create
employment, and that tourist expenditure would generate a large multiplier effect that
would stimulate the local economy (Telfer, 2002). In some cases the large-scale development of tourism was initiated by state investment, for example the Mexican strategy
of building a number of very large resort complexes in such places as Cancun in the
hope that economic benefits would accrue to a larger area. Consequently, tourism as a
form of modernisation involved the transfer of capital, technology, expertise and
'modern' values from the West to less-developed countries.

In the 1970s, as tourism increased as an economic activity, its benefits were
questioned as economic studies pointed to lower multiplier effects and higher levels of

Think point

How did tourism manifest itself in the modernisation process during the 1960s?

economic leakage than were expected. The application of the social science disciplines of anthropology and sociology to tourism studies also emphasised the negative social and cultural impacts that could be caused by tourism development. According to Telfer (2002), these criticisms were similar to the dependency theorist critique of modernisation, emphasising growing uncertainty of using tourism as a development tool.

In the 1970s, within the general context of dependency theory, the emerging criticism of tourism as perpetuating exploitative economic relationships between metropolitan generating countries and peripheral destination societies became evident. Turner and Ash (1975) referred to tourism destinations as the pleasure-periphery, expressed geographically as the tourist belt that surrounds the industrialised zones of the world, for example the relationship of the coastline of the Mediterranean Basin with Northern Europe and that of the islands of the Caribbean with the US. The corollary of tourism as a new type of plantation economy also emerged as the exploitative relationship between the developed and developing world was emphasised.

Terms equating tourism with 'neo-colonialism' and 'imperialism' also entered the vernacular. Hall (1994) suggests that these terms are powerful metaphors to describe the relationship between core and periphery areas, illustrating the potential loss of control that the host community may experience in the face of foreign tourism interests and the actions of local elites. Similarly, in the view of Britton (1982), when a developing country uses tourism as a development strategy it enters a global system over which it has little control. Consequently, as the international tourism industry is a product of metropolitan capitalist enterprise, possessing the superior entrepreneurial skills, resources and commercial power that characterise metropolitan countries, it is thus able to dominate the tourism economies of many developing-world destinations.

Dependency theorists also point out that the inherent social and structural frameworks that exist within a society determine the way the international tourism industry integrates with it (Britton, 1982). In the dependency scenario of international tourism, a model of multinational companies working with elites replaces the model of colonial governments working with local elites. Thus, only the privileged commercial and political groups in the periphery, along with foreign interests, are in a position to coordinate, construct, operate and profit from the development of a new industry such as tourism. It is consequently argued that the economic outcome of such a model is the removal of part of the economic surplus by foreigners, and the non-productive use of much of the remaining surplus by ruling elites.

The role of multinationals operating in the tourism industry is likely to have an influential effect upon the economic outcomes of tourism development. Certainly as tourism has increased in volume during the latter half of the twentieth century, and opportunities have been created for increased profitability, there has been a maturity of the tourism industry. This has meant an increase in the size of key operators and the allowance of

more market influence and control, as was explained in Chapter 4 in reference to monopolies and oligopolies. The role of tourism multinationals in the development of tourism
in peripheral areas is underlined by Britton (1982: 336):

> The establishment of an international tourist industry in a peripheral economy
> will not occur from evolutionary, organic processes within that economy, but
> from demand from overseas tourists and new foreign company investment, or
> from the extension of foreign interests already present in that country.

The dependency of the peripheral upon the core is further increased by the need for
air transport to move tourists to destinations. A key mechanism for metropolitan firms
to control tourist flows is through international transport, especially so for islands.
The dependency of a peripheral area on a core is exemplified through the case of
Fiji in Box 5.3.

Think point

Summarise the main criticisms that dependency theorists have made of the international
tourism system. Do you think they are valid?

Box 5.3 Tourism in the periphery, the case of Fiji

In a seminal detailed study of Fiji, Britton (1982) highlighted the dependency that
peripheral areas can have upon core areas. As an island of the Polynesian archipelago, Fiji has used international tourism to develop its economy. However, it is
heavily dependent upon foreign airlines to bring tourists to the island. In 1978,
Quantas and Air New Zealand were responsible for 80 per cent of the airline seats
to Fiji. The ability of regional carriers, such as Air Pacific or Polynesian Airways,
to compete with these airlines was negated by high operational costs, limited equipment capacity, and interference by foreign management and shareholding interests.
Direct pressure by metropolitan governments also ensured protection for their
national airlines and carriers.

The case of Fiji also illustrates the problems faced by a country or region that has
a limited manufacturing base. Although Fiji is in fact one of the largest Pacific Island
economies, it relies upon 53 per cent of hotel food purchases; 68 per cent of standard hotel construction; and 95 per cent of tourist-shop wares to be supplied by
imports. The granting of management contracts for hotels to foreign companies
further extends the influence of foreign control. This gives effective corporate control
to foreign companies without them having to commit large sums of money in capital
investment. The consequence of this heavy dependence upon foreign interests is that
over 70 per cent of tourist expenditure is lost to pay for imports, profit expropria-
tion and expatriate salaries.

Source: Britton (1982)

However, similar to general criticisms of dependency theory, the application of dependency theory to tourism does not offer solutions of how to maximise local economic benefits nor reduce the negative environmental and social impacts. To attempt to find solutions to these issues it is necessary to consider the alternative development paradigm in the context of tourism, the most popular manifestation of this being sustainable tourism development. Given that the Brundtland Report was concerned with issues of environmental degradation; poverty; gender equality; democracy; human rights; and intra- and inter-generational equity, it could be expected that sustainable tourism would emphasise these different components. However, the concept of sustainable tourism has a similar ambiguity inherent to the concept of sustainable development, with no agreed common definition.

The ambiguity of the concept of sustainability means that the political context, and especially the political values of those who have power and decision-making, will be influential in determining the interpretation of sustainable tourism. Butler (1998) points out that it is not possible to separate sustainable tourism from the value systems of those involved and the societies in which they exist. Concerning this latter point, Mowforth and Munt (1998: 122) remark:

> If it remains a 'buzzword' which can be so widely interpreted that people of very different outlooks on a given issue can use it to support their cause, then it will suffer the same distortions to which older-established words such as 'freedom' and 'democracy' are subjected.

They continue to demand that if sustainable tourism is really to be achieved then there is a 'need to politicise the tourism industry in order to promote its movement towards sustainability and away from its tendency to dominate, corrupt and transform nature, culture and society' (Mowforth and Munt, 1998: 123). House (1997) also infers a political dimension to the application of sustainability in tourism. She recognises two polarised positions or 'schools of thought' reflecting different ideologies on the employment of the concept. At one end are the 'reformists', whose ideology and actions of implementation are preoccupied with the status quo and reluctant to challenge the existing social, political and economic structures that underpin tourism development (1997: 93). Conversely, 'structuralists' possess a much more radical view of tourism development which challenges the paradigm on which economic, social and political development is based. The structuralist model of tourism is therefore one that is radical, involving questioning the values of society, as much as those of tourism.

It is therefore necessary to realise that sustainable tourism is not solely connected with the conservation of the physical environment but incorporates cultural, economic and political dimensions. As has been pointed out, the ambiguity of the term means that it can be interpreted and owned by so many different groups with opposing ideologies that trying to agree a common definition of the term is meaningless. Perhaps the most useful way of thinking about sustainability is not necessarily to think of it as an end point, but to think of it more as a guiding philosophy which incorporates certain principles concerning our interaction with the environment. The next section of this chapter considers the guiding principles that have been developed in connection with tourism.

> **Think point**
>
> Carry out an Internet search of the terms 'sustainable development' and 'sustainable tourism'. Who uses these terms? Do they have different economic and political interests?

Application of sustainability in tourism

The concept of sustainability has been applied in the tourism sector in different ways, at both national and local levels, and in the public, private and voluntary sectors. The last decade of the twentieth century saw an increased effort by some private-sector tourism organisations, to make it evident that they were placing the environment in a more central position to their operations and attempting to become more 'sustainable'. The extent to which this is from a genuine concern for the environment, or a business ploy to attract more customers and an attempt to stave off regulation of the industry, is unsure. Butler (1998: 27) suggests that the tourism industry has adopted sustainability for three reasons: 'economics, public relations and marketing'.

However, initiatives have been taken in partnership between the private sector and other organisations. One example of a scheme that has been launched by the tourism industry with the support of the United Nations Environment Programme (UNEP), the United Nations Educational, Scientific and Cultural Organisation (UNESCO) and the World Tourism Organisation (WTO) is the Tour Operator's Initiative for Responsible Tourism, as described in Box 5.4.

Nevertheless, it is not just the tourism industry that has taken the initiative to develop a more sustainable form of tourism. An example of an initiative to improve the environmental quality of tourism taken by an NGO, the 'Foundation for Environmental Education', is described in Box 5.5.

One of the first public strategies on tourism and sustainability emerged from the Globe '90 conference in Canada, which brought together government, NGOs, the tourism industry and academics to discuss the future relationship of tourism with the environment. Five main goals of sustainable tourism were identified:

> (1) to develop greater awareness and understanding of the significant contributions that tourism can make to the environment and economy; (2) to promote equity and development; (3) to improve the quality of life of the host community; (4) to provide a high quality of experience for the visitor; and (5) to maintain the quality of the environment on which the foregoing objectives depend.
>
> (Fennell, 1999)

As with the concept of sustainability, the goals tend to be all-encompassing, potentially conflicting and give little guidance on how tourism should be developed.

Nevertheless, those who favour increased local democracy, notwithstanding the problems of defining the meaning of 'community' and the inherent tensions and

Box 5.4 Tour Operator's Initiative and responsible tourism

One example of a scheme that has been launched by the tourism industry with the support of the United Nations Environment Programme (UNEP), the United Nations Educational, Scientific and Cultural Organisation (UNESCO) and the World Tourism Organisation (WTO) is the Tour Operator's Initiative for Sustainable Tourism Development, launched on 12 March 2000. The Initiative has three main objectives:

1   'Assist tour operators to put sustainable tourism into practice'. Emphasis is placed upon: the pursuit of best current practice in their internal operations; supply chain management; and relations with customers and destinations.
2   'Broaden support for sustainable development amongst other players in the tourism industry – including tourists'. Emphasis is placed upon cooperation with the other stakeholders in tourism, including other business partners, government, local communities and NGOs.
3   'Create a critical mass of committed tour operators'. Emphasis is placed upon raising the image of committed tour operators and the image of the Initiative as a world leader in environmentally, socially and culturally responsible tourism.

To become a member of the Initiative the company's senior management must make a corporate commitment to sustainability by signing the Initiative's Statement of Commitment. The specific commitments are to:

■   protect the natural environment and cultural heritage;
■   cooperate with local communities and people, ensure they benefit from the visits of our customers and encourage our customers to respect the local way of life;
■   conserve plants and animals, protected areas and landscapes;
■   respect the integrity of local cultures and their social institutions;
■   comply with local, national and international laws and regulations;
■   oppose and actively discourage illegal, abusive or exploitative forms of tourism;
■   provide information on our activities to develop and encourage the sustainable development and management of tourism;
■   communicate our progress in implementing this commitment.

(UNEP, 2002)

The founding members of the Initiative include large mass-tourism operators including TUI (Germany), Accor Tours (France) and First Choice (UK). The extent to which the scheme proves to be successful can only be determined in the long term but its symbolism is representative of a growing awareness in the industry to be more environmentally and socially accountable, and to demonstrate corporate social responsibility (CSR).

Box 5.5 Blue Flag campaign

Blue Flag is a campaign run by the independent non-profit organisation 'Foundation for Environmental Education'. Its aim is to work towards sustainable development in coastal areas used for tourism by raising levels of water quality, environmental management, environmental education and information, and safety. The scheme was launched by the Foundation for Environmental Education in Europe (FEEE), as one of several 'European Year of the Environment activities' in 1987, with 244 beaches and 208 marinas from 10 different countries being awarded the Blue Flag. In 2001, FEEE decided to become a global organisation, changing its name to Foundation for Environmental Education (FEE). The spreading of the campaign to other regions of the world is supported by the United Nations Environment Programme (UNEP) and the World Tourism Organisation (WTO). By 2004, 2,312 beaches and 605 marinas had been awarded the Blue Flag in 25 different countries, most of them in Europe, but also including Montenegro and South Africa. Five Caribbean countries, the Bahamas, Jamaica, Barbados, and the Dominican Republic are to pilot schemes. Included within the criteria to be awarded Blue Flag status are that: no industrial or sewage related discharges may affect the beach area; no algal or other vegetation may accumulate and be left to decay on the beach; information on natural sensitive areas in the coastal zone, including its fauna and flora, must be publicly displayed and included in tourist information; the local community has an Environmental Interpretation Centre dealing with the coastal environment; the local community must have a land-use and development plan for its coastal zone; beach guards are on duty during the bathing season; and first aid must be readily available on the beach.

Source: Blue Flag (2004) http://www.blueflag.org

divisions within it, have often advocated community control over development decision-making. A more radical perspective of sustainable development supports development decision-making at local level, not only on democratic principles, but also on the presumption that local people are more likely to act as stewards of the environment than external parties. However, community participation in planning and development may or may not be successful in encouraging people to favour less environmentally damaging development options, as attitudes to the physical environment are likely to reflect economic priorities, as discussed in Box 5.6.

To realistically encourage stewardship of the environment by local communities, forms of tourism will need to be developed that are not only sympathetic to the environment, but also offer economic benefits as is the case in Senegal, described in Box 5.7.

Think point

How is the success of development evaluated in the case study of Senegal? What factors have made tourism successful as a means for development in this case?

As the case of Senegal suggests, alternative tourism is likely to be smaller in scale and have a higher degree of community involvement than mass tourism. While there is no universal agreement on a definition of what alternative tourism actually is (Brown, 1998), the differences of alternative tourism to mass tourism are highlighted by Cater (1993: 85) as: 'Activities are likely to be small scale, locally owned with consequentially low impact, leakages and a high proportion of profits retained locally. These contrast with large-scale multinational concerns typified by high leakages which characterise mass tourism.' Other aspects of alternative tourism are a reflection of the alternative development agenda, including issues of women's empowerment and democracy. The characteristics of alternative tourism that differentiate it from mainstream tourism are shown in Box 5.8.

A kind of tourism that is often held to typify alternative tourism is 'ecotourism'. Like the concept of sustainable tourism it is popular with a variety of stakeholders, owing to a

---

Box 5.6 Community decision-making and tourism

It is often assumed that advocating community development decision-making in tourism is likely to encourage the conservation of natural resources. However, this principle is not always the case. The local community may not favour tourism development even when it is presented as a less environmentally damaging development option than other forms of economic activity. For example, Burns and Holden (1995) comment upon the case of tourism development in the St Lucia Wetlands in Natal, South Africa, an area containing coral reefs, turtle beaches, highly afforested dunes, freshwater swamps, grasslands and estuaries. Rio Tinto Zinc (RTZ), the giant transnational mining corporation, wanted to mine the dunes for titanium dioxide slag. Despite assurances from RTZ over redressive environmental restoration of the area when the mining had ceased, central government was opposed to the use of the area for this purpose on environmental grounds, and instead favoured the development of ecotourism. However, local people, mainly Zulus, favoured the development of mining on the basis that RTZ had a good track record of paying comparatively high wages and investing in schools, clinics and other facilities. The Natal Parks Board, who run the surrounding game parks, were perceived by the local community as paying low wages, and having displaced local people from their lands to establish game reserves in the 1960s and 1970s.

Similarly at Cairngorm, in the Scottish Highlands of Britain during the 1990s, there was a high level of controversy over the planned development of a funicular railway up the mountainside for the purposes of downhill skiing. Opposition to the scheme from major NGOs such as the World Wildlife Fund (WWF) and the Royal Society for the Protection of Birds (RSPB) was based upon the possible environmental impacts on the arctic-alpine environment, which is unique within the British Isles. However, instead of receiving the support of the majority of local people, the WWF and the RSPB were largely seen as outsiders attempting to stop economic development to protect birdlife and flora, thereby denying local people employment and other economic opportunities.

similar ambiguity in its meaning (Wheeller, 1993). While it is generally agreed upon that ecotourism has nature as its central focus, there is a more general confusion about its political and economic dimensions as is highlighted by Cater (1994: 5): 'In particular, the term ecotourism is surrounded by confusion. Is it a form of "alternative tourism" (furthermore,

---

Box 5.7 Integrated rural tourism, Lower Casamance, Senegal

The development of tourism in the Lower Casamance is an example of how tourism can be used as a tool to enrich the livelihood and well-being of rural peoples. The aim of the scheme was to aid development, and also provide a more meaningful interchange between the local people and tourists than was being experienced on the coast, where hotels had largely been built with foreign capital and local people were excluded from tourist complexes by security guards and high walls.

In total 13 tourist camps have been built, from an initial investment of $7,000 each, provided by l'Agence de Cooperation Culturelle et Technique. Tourists stay in simple lodges, built using traditional materials in the local architectural styles, so diminishing the differentiation between tourist and local facilities. Tourist numbers are restricted to a maximum of 20–40 guests and lodges are only constructed in villages where the population is 1,000 or above. Tourists eat locally grown produce wherever possible following traditional recipes. The scheme has proved to be a success, aiding development and social stability, improving health and educational facilities and, critically, providing employment opportunities for the young which discourages them from migrating to larger towns to look for employment. Public expenditure of the revenues from tourism is controlled by village cooperatives.

Source: Gningue (1993)

Box 5.8 Characteristics of alternative tourism

- pace of development directed and controlled by local people rather than external influences;
- small-scale development with high rates of local ownership;
- environmental conservation and the minimisation of negative social and cultural impacts;
- maximised linkages to other sectors of the local economy, such as agriculture, reducing a reliance upon imports;
- maximisation and an equitable distribution of the economic benefits of tourism for local people;
- empowerment of women and other marginalised groups in democracy and decision making;
- attracting a market segment that is willing to accept local standards of accommodation and food and that is interested in education in the local culture and environment.

Think point

How do the characteristics of alternative tourism contrast with those that might typify neoliberal types of tourism? While local democracy and community participation may sound good in theory, should this principle be encouraged if the local community favours tourism that has destructive consequences for nature?

---

what is "alternative tourism"?)? Is it responsible (defined in terms of environmental, socio-cultural, moral or practical terms)? Is it sustainable (however defined)?'

The above quotation raises key questions about ecotourism, relating to its political and economic structure, environmental and social responsibility, and its long-term sustainability. For instance, is ecotourism about challenging the political and economic influence of multinational hotel corporations and operators; is ecotourism about empowering local communities with the rights to determine development decision-making, thereby challenging the influences of central and regional governments; is it a form of more balanced and environmentally sustainable development; or is it purely about promoting the nature and physical characteristics of environments to encourage more tourism?

Given the variance in how ecotourism can be interpreted, it is unsurprising that it has manifested itself in different ways, as is described in Box 5.9 with reference to the case studies of Belize and Bolivia.

Box 5.9 Two different types of ecotourism

The case of Belize

The first country to promote itself as a major ecotourism destination was Belize in Central America. The attraction to the government of the development of ecotourism was to earn foreign exchange. The government committed itself to balanced and environmentally sound tourism, after the 1989 general election, under the slogan of 'Tourism with Dignity'. The typical advertising slogans used to promote Belize are 'Belize so natural'; 'friendly and unspoilt'; and 'naturally yours'. The development of ecotourism in Belize is based upon its natural resources, which include a diverse range of vegetation types, notably mangrove swamps, wetland savannah, mountain pine-forests and tropical rainforests. Additionally, the coral reef off Belize's coast is the second longest in the world after the Australian Barrier Reef, and therefore represents a major natural attraction in its own right. There are also several notable archaeological sites of the Mayan civilisation.

Yet, despite the government's statements to develop a form of tourism with dignity, the 'ecotourism' that has begun to manifest itself bears many of the political, economic and environmental hallmarks of mass tourism. Development has taken place within a free-market environment, which has encouraged unregulated entrepreneurship. The financial opportunities to be gained from ecotourism have led to rapid property inflation and it is now calculated that 90 per cent of all coastal development is in foreign hands. US developers are building luxury resorts and golf courses and it is estimated that 65 per cent of the members of the Belize Tourism Industry Association are expatriates from the US. It is evident that the type of tourism that is developing in Belize is displaying a high degree of external foreign ownership vis-à-vis a more bottom-up, local community-based approach to tourism development. Similarly, although the majority of tourists who come to Belize do not arrive on package tours, their travel and accommodation is organised by foreign tour operators, and subsequently the economic leakage factor is likely to be high.

The scale of development associated with ecotourism is also causing environmental problems. Some of the coral in the Hol Chan Marine Reserve established in 1987 has been broken from diving and tourist visitation. There has also been a decline in the numbers of conch and lobster, owing to over-fishing, partly to satisfy tourism demands. Other examples of the impacts of ecotourism are more dramatic. In 1992, 'Eco-terrorism at Hatchet Cay' was the headline in a local paper; in reference to a US resort owner who had attempted to blow up part of the coral reef to make his resort more accessible to visiting boats.

Sources: Cater (1992); Mowforth and Munt (1998); Panos (1995)

The case of Bolivia's Noel Kempff Mercado National Park

The Noel Kempff Mercado National Park covers approximately 15,000 square kilometres of rainforest, savannah and other ecosystems on Bolivia's border with Brazil. It is subsequently very biodiverse in flora and fauna, including more than 4,000 plant species, endangered animals, including the jaguar, tapir, rhea and giant anteater, as well as being home to 630 bird species. During the 1970s and 1980s, the local economy was based on logging and hunting, with countless tons of mahogany and tens of thousands of jaguar and other animal hides being exported from the area, before its designation as a national park.

In an attempt to conserve the natural resources, ecotourism has now been established, in cooperation with local villages situated on the borders of the park. Villagers have been trained as guides, and granted financial loans through the Global Environment Facility/Small Grants programme (SGP), supported by the United Nations Development Programme (UNDP). The loans are used to buy hiking and camping equipment to take groups of tourists into the park, who pay a daily fee of US$10 to local guides. A rotating credit fund is also available through the SGP, which helps families to build rustic accommodation, equip small restaurants, and buy bicycles and dugout canoes to rent to tourists. The project is consequently providing people with economic opportunities and empowering women while helping to ensure conservation, which is achieved through the careful environmental management of tourism by the park authorities in consultation with local people.

Source: Dudenhoefer (2002)

It would seem that ecotourism is manifesting itself in two main ways. In the first model, ecotourism is being used to meet economic objectives by promoting the quality of the environment to attract international tourists. Ecotourism is incorporated into a global market, in which development is based upon foreign inward investment, with little government interference in terms of planning and environmental regulation. In environmental and social terms such a model would be unlikely to be sustainable, and as tourism development progressively destroys the quality of the resources tourists are coming to see, it would also prove to be economically unsustainable.

In the second scenario, ecotourism is developed at a local level, and emphasis is placed upon resource conservation financed through the revenues from international tourism. Ecotourism is characterised by being low-scale and low-impact, with a high degree of local involvement in development decision-making, and a high level of local ownership of tourism facilities. In this scenario the influence of multinational corporations is minimised, although there would still be a reliance on international intermediaries such as foreign airlines to transport tourists, and niche tour operators to promote the destination.

Think point

Considering the characteristics of alternative tourism highlighted in Box 5.8, which of the models of ecotourism described in Box 5.9, could be best described as 'alternative tourism'?

Despite the ambiguity inherent to the meaning of sustainable tourism and ecotourism, they represent important ideals for international policy on tourism, as is exemplified by the declaration of the United Nations highlighted in Box 5.10 about tourism's role in sustainable development.

Think point

Referring to the United Nations statement on sustainable tourism development in Box 5.10, do you think this favours a radical approach to tourism development or one that is more in line with the principles of economic neoliberalism?

Pro-poor tourism/sustainable tourism – eliminating poverty

The comparative advantage that lesser developed countries have in terms of their natural and cultural resources for tourism, combined with the limited development opportunities in other sectors of the economy, has meant that the role of tourism as a means to combat poverty is increasingly being encouraged by the World Tourism Organisation. This initiative was taken at the World Summit on Sustainable Development in Johannesburg in 2002, being given the title: 'Sustainable Tourism as an effective tool

---

Box 5.10 *Report of the World Summit on Sustainable Development*, United Nations, New York

43. Promote sustainable tourism development, including non-consumptive and eco-tourism, taking into account the spirit of the International Year of Eco-tourism 2002, the United Nations Year for Cultural Heritage in 2002, the World Eco-tourism Summit 2002 and its Quebec Declaration, and the Global Code of Ethics for Tourism as adopted by the World Tourism Organization in order to increase the benefits from tourism resources for the population in host communities while maintaining the cultural and environmental integrity of the host communities and enhancing the protection of ecologically sensitive areas and natural heritages. Promote sustainable tourism development and capacity-building in order to contribute to the strengthening of rural and local communities. This would include actions at all levels to:

(a) Enhance international cooperation, foreign direct investment and partnerships with both private and public sectors, at all levels;

(b) Develop programmes, including education and training programmes, that encourage people to participate in eco-tourism, enable indigenous and local communities to develop and benefit from eco-tourism, and enhance stakeholder cooperation in tourism development and heritage preservation, in order to improve the protection of the environment, natural resources and cultural heritage;

(c) Provide technical assistance to developing countries and countries with economies in transition to support sustainable tourism business development and investment and tourism awareness programmes, to improve domestic tourism, and to stimulate entrepreneurial development;

(d) Assist host communities in managing visits to their tourism attractions for their maximum benefit, while ensuring the least negative impacts on and risks for their traditions, culture and environment, with the support of the World Tourism Organization and other relevant organizations;

(e) Promote the diversification of economic activities, including through the facilitation of access to markets.

UN (2002: 33)

---

for Eliminating Poverty' (ST-EP), building on work that had already been initiated by the Department for International Development (DFID) in the UK under the aegis of 'Pro-Poor Tourism' (PPT) (Sofield *et al.*, 2004).

Although there are many definitions of poverty, half of the world's population live on less than US$2 per day, as highlighted in Box 5.2, virtually all of whom live in the less-developed countries. Hence the focus of 'pro-poor tourism' is upon tourism destinations in the South and developing tourism good practices that are particularly relevant to the condition of poverty, being defined thus by Ashley *et al.* (2001: 2): 'Pro-poor tourism is defined as tourism that generates net benefits for the poor.' The distinctive approach about ST-EP and PPT is that it puts poor people and poverty at its centre, distinguishing it from sustainable development, which emphasises more general development based upon sustainable principles. A key aspect of this focus is that in some cases people have ownership of resources such as cultural festivals and wildlife and scenery, which may be utilised for tourism development (Sofield *et al.*, 2004).

The basis of pro-poor tourism is therefore about the empowerment of the poor and benefiting their livelihoods, which include environmental, cultural and social benefits besides purely economic ones. The means for achieving this is to give the poor access to tourism markets and provide them with the resources to participate in tourism. The advantage of tourism as an economic activity is that the customers come to the destination, it is relatively labour intensive, and it employs a high proportion of women (WTO, 2002). One of the major problems faced by many poor people in the world who may wish to participate in tourism is that they are denied access to tourists. This is particularly so when the movements of tourists are controlled by tour operators and companies, for example when tourists purchase 'all-inclusive' holidays and spend most or their time in hotel resorts and complexes; or on cruise ships; in coaches and safari vehicles; and inside sites and managed attractions. For there to be a meaningful interaction between the poor and tourists there needs to be the provision of some kind of interface between the tourists and local people, to enable them to sell handicrafts or other types of products and services to the tourists. The emphasis of pro-poor tourism is therefore upon unlocking opportunities for the poor within the tourism sector.

One way of combating this is for partnerships of hotels and tour operators to work together to encourage local people to develop tourism products and services and support them with training and marketing. Examples from Africa of the sorts of products and services that could potentially be offered to tourists by the poor include: drumming and dance classes; story telling; sharing a meal with a family in a village; hair braiding; bird watching; school visits; history tours; and guided walks to look at plants and medicinal herbs (WTO, ibid.). The emphasis is therefore for communities to engage in 'complementary products', that is, to build upon the existing mainstream tourism businesses and infrastructure. As a form of additional income it can be important not only in terms of lifting people above the threshold of poverty, but also the extra income gained through tourism can support a variety of other family members beside the direct recipient of the income.

The targeting of the poor specifically for development is a recognition that development strategies based upon a top-down model and emphasising the trickle-down effect have not worked. As the WTO (2002: 59) comment:

> It is now more widely recognised that a significant reduction in poverty can be achieved only if the benefits of growth are redistributed to the poor or if the poor themselves can be brought into economic activity either through employment or entrepreneurial success.

The emphasis upon 'redistribution' of the benefits would therefore seem to indicate a change in the relative benefits of tourism between different groups in society.

Think point

How can tourism be used to combat poverty?

Summary

- Economic theory will not determine that the economic benefits of tourism are equally distributed between countries or people. Resource distribution is not purely a question of economics but is also dependent upon power relationships and politics. To understand the role tourism plays in development it is subsequently necessary to understand the inter-relationship between economics and political processes.

- It was particularly in the immediate post-Second World War period that a strong association was made between economic progress, development and politics. Different paradigms of development have been formulated, including modernisation; dependency; economic neoliberalism; and alternative/sustainable. All contain differing political and economic emphases, which in turn influence how tourism will be used as a means to development.

- The World Bank and the United Nations from the late 1950s supported tourism as a means for development. This coincided with the beginnings of mass participation in international tourism by the populations of the developed countries. In the 1960s, tourism was largely seen as a means of modernising less-developed countries, involving the transfer of capital, technology and expertise from western countries. During the 1970s, the international tourism system was criticised by dependency theorists as a perpetuation of exploitative economic relations between the 'core' of developed countries and the 'periphery' of less-developed countries. Within economic neoliberalisation emphasis is placed upon tourism as an export and the opening up of a country's national and cultural resources to foreign investment and control.

- The failure of traditional approaches to development has led to the emergence of an alternative agenda for development of which sustainable tourism and ecotourism are key parts. The terms have become popular ones among different stakeholders, including international agencies, governments, private industry and NGOs. However, the willingness of organisations with different and sometimes opposing political and economic interests to adopt these terms is reflective of their ambiguity. Do they represent simply new types of tourism within the existing framework of global society, or are they representative of a questioning of the values of a society, and a subsequent call for its political and economic restructuring?

- Increasing attention is being turned to tourism's role in combating poverty in less-developed countries. The difference between pro-poor tourism (PPT) and sustainable tourism is that poor people and poverty are the central focus of PPT, whereas sustainable tourism emphasises more general development based upon sustainable principles. A key focus of PPT is the empowerment of the poor, providing the resources for their involvement in tourism, and, critically, access to markets from which they may be presently excluded.

Suggested reading

Fennell, D. (1999) *Ecotourism: An Introduction*, Routledge, London.

Harrison, D. (1988) *The Sociology of Modernisation and Development*, Unwin Hyman, reprinted (1991) by Routledge, London.

Harrison, D. (ed.) (2001) *Tourism and the Less Developed World: Issues and Case Studies*, CABI, Wallingford.

Lea, J.P. (1988) *Tourism and Development in the Third World*, Routledge, London.

Sharpley, R. and Telfer, D. (eds) (2002) *Tourism and Development: Concepts and Issues*, Channel View Publications, Clevedon.

Turner, L. and Ash, J. (1975) *The Golden Hordes: International Tourism and the Pleasure Periphery*, Constable, London.

Suggested websites

New Internationalist   www.newint.org
The World Bank   www.worldbank.org
United Nations   www.un.org

ANTHROPOLOGY AND TOURISM

6

This chapter will:

- introduce anthropology as a discipline;
- outline anthropological perspectives on tourism;
- consider how tourism can be viewed as a search for the authentic;
- critically discuss tourism as a form of ritual;
- evaluate cultural change as a consequence of tourism;
- describe strategies for the protection of culture.

Introduction

Similar to many of the social sciences, the roots of anthropology rest in the changing economic and social conditions of the nineteenth century. A combination of an industrialising and changing society, and European expansion led to the development of anthropology as an academic subject. European colonial expansion created the means for wider contact between societies, offering anthropologists the opportunity to work in colonial areas, to undertake studies of social organisation, customs and religions (McLeish, 1993). Anthropologists were therefore offered a view of the workings of pre-industrial societies that contrasted with the complexity of modern industrial western societies. Early anthropologists, such as Edward Tylor, were interested in tracing everything from writing systems to the marriage practices of cultures outside Europe. However, although the origins of anthropology lay in the desire to record ways of life in small-scale technologically simple societies, it would be a mistake to assume that these societies were unchanging or always even truly isolated before their contact with the West.

As a definitive strand of the social sciences, the two major figures associated with the founding of anthropology as a social science discipline include Sir James Frazier (1854–1951) and Bromslow Malinowski (1884–1942). However, Frazier and Malinowski had different approaches to social anthropology as an area of study. Frazier was interested in trying to discover fundamental truths about the nature of human psychology, through a detailed comparison of human cultures on a world scale, even though he had no first-hand acquaintance with the peoples he wrote about (Leach, 1996).

By contrast, the approach of Malinowski was to emphasise how 'primitive' communities functioned as a social system and how individual members passed through their lives within them. Malinowski's analysis was based upon field research he had personally conducted over a period of four years in a small village in Melanesia.

It is the latter approach, emphasising a prolonged period of field research, that has helped define social anthropology as a distinctive discipline. The anthropological approach to understanding cultures outside Europe, which have often been referred to as 'primitive' cultures, is heavily based upon Malinowski's methodology termed 'ethnography' (McLeish, 1993). Malinowski advocated that anthropologists engage in extended fieldwork, spending a much longer duration of time living with the community than had been the norm up to then, a method he called 'participant observation'. Ethnography is based on the apparently simple idea that in order to understand what people are up to, it is best to observe them by interacting with them intimately over an extended period, usually of several years. This is why anthropologists have traditionally spent long periods, sometimes years at a stretch, living in the communities they study.

The importance of ethnography to anthropology is underlined by Monaghan and Just (2000), who emphasise that it is a key aspect of what defines anthropology from the other social sciences. They comment:

> As has often been said, if you want to understand what anthropology is, look at what anthropologists do. Above all else, what anthropologists do is ethnography. Ethnography is to the cultural or social anthropologist what lab research is to the biologist, what archival research is to the historian, or what survey research is to the sociologist.
>
> (2000: 13)

So although anthropologists' interests may overlap with other disciplines, notably sociology, anthropology has developed a distinctive approach to its subject matter. The distinctiveness and importance of ethnography to anthropology is also emphasised by Nash (1996), who suggests that an anthropologist is required to live with people on intimate terms while getting to know them, and that a further hallmark of the ethnographer is that they conduct their fieldwork in the language of the culture without the use of a translator.

The advantage of ethnography is that by spending time in another cultural environment and witnessing the social and environmental context of human action, the social anthropologist can open nuances of understanding that may not be apparent through the application of research techniques of other disciplines. Through establishing long-term relations with people and developing friendships, the ethnographer is subsequently able to ask questions and gain responses that pass beyond the superficial and have a cultural context. Smith and Brent (2001) suggest that such a holistic understanding of human society, combined with its methodology of cross-cultural analysis, is the hallmark of anthropology. However, as Nash (1996) points out, ethnography is no longer reliant solely on interviews, dialogues and observations, now employing a range of more specialised techniques including audio recording, video, drawing and mapping.

Think point

What are the characteristics of ethnography as a research method?

Although anthropological enquiry traditionally emphasised the understanding of societies through their structures and social relationships, a major break with this school of thought came in the 1960s based on the work of the French anthropologist Claude Levi-Strauss. In place of analysing culture through the structures of society, he emphasised the conceptual structures of the mind, such as myths, symbols and totemism. Additionally, in the latter part of the twentieth century anthropology moved away from an exclusive concern with non-western small-scale rural societies to embrace other various social contacts. These include the anthropologist's own societies, and groups that previously would have been the prerogative of sociology such as labour unions, social clubs and migrant communities (McLeish, 1993; Nash, 1996). The diversity of what may be studied within the realm of anthropology is reflected in the different areas that now command anthropological attention, including development, tourism, visual anthropology, emotions and ethnicity (McLeish, 1993). In the case of tourism, it is the field of sociocultural anthropology, which focuses on peoples' behaviour, that is most relevant (Nash, 1996).

A key concern of anthropologists lies with 'culture', usually used in an anthropological sense synonymously with human groups or societies. Yet, the meaning of the term of culture is contestable. A common interpretation of culture dating to Victorian times is something that a nation, class or group of people might possess that differentiates them from others. For example, the person who goes to the opera, sips champagne and reads Proust may be regarded as being more cultured than one who goes to a soccer match, drinks beer and reads pulp fiction. However, the interpretation of the meaning of 'culture' to infer a kind of social superiority has been rejected by anthropologists. In reference to Britain, though applicable to many westernised countries, Franklin (2003) notes that prior to the 1960s, art history, literature, music and many other cultural forms were differentiated by terms of high or low culture. Nevertheless, although this differentiation may be less marked in contemporary society, it is still evident and continues to exist.

However, despite there being more anthropological definitions of culture than there have been anthropologists, Monaghan and Just (2000: 35) comment:

> However we define culture, most anthropologists agree that it has to do with those aspects of human cognition and activity that are derived from what we learn as members of society, keeping in mind that one learns a great deal that one is never explicitly taught.

In a similar vein, Smith (2003: 9) suggests: 'Hence culture is viewed as being about the whole way of life of a particular group or social group with distinctive signifying systems, involving all forms of social activity, and artistic or intellectual activities.' Accepting such a definition of culture enables a wider view of what it encompasses than it being defined only by the behaviour and types of activities that a particular social class participates in. As she points out, the advantage of such a definition is that it can transcend the barriers of individual and group culture, while at the same time conveying the importance of both heritage and tradition, beside that of contemporary culture and lifestyles.

Think point

How would you define culture?

Anthropology and tourism

The application of anthropology to tourism is associated with the growth in international tourism in the second half of the twentieth century. Particularly, the increase in tourism to the less-developed world meant that tourists were visiting countries in which many anthropologists had carried out their fieldwork (Nash, 1996). The consequent contact and interaction of different cultures on a scale never witnessed before lends itself to an anthropological approach. In terms of the founding of the anthropology of tourism as a field of academic enquiry, Smith and Brent (2001) date it to 1974, at the American Anthropological Association meeting held in Mexico City.

Based upon the writings of Crick (1988), Selwyn (1996) identifies three main overlapping strands of enquiry within the anthropology of tourism: social and cultural change; semiology of tourism; and tourism's political economy. The approach taken within this chapter is to integrate these broad themes into the two major foci of the anthropological investigation of tourism, i.e. the tourist and the tourist destination. This broad but interrelated division provides a framework for the exploration of key aspects of tourism and anthropology, including mythology, ritual, authenticity, development and cultural change.

The emphasis in the anthropological enquiry of these issues centres upon the use of ethnography as a methodology. To reiterate the advantage of this technique compared to other methodologies it is that it permits the possibility of understanding the nuances of tourism that otherwise might be missed or incomprehensible. Subsequently, it can be argued that based upon the use of ethnographic methodology and a focus on culture, anthropologists have special preoccupations and points of view in their approach to tourism. The chapter now proceeds to examine how these particular preoccupations have shaped the anthropological understanding of tourism.

The tourist

Although most of the anthropological research in tourism has focused on tourism's effects on peoples in less-developed countries, this section commences with the anthropological study of tourists, as it is the tourist who potentially represents the principal proponent or actor of cultural change. Also, focusing on the tourist partly responds to Nash's (1996) concerns over the little emphasis that has been placed within anthropology on the conditions that generate tourists and tourism. To this consideration it may be added that there is also an absence of anthropological research into how, in turn, tourists affect the cultures that they are returning to. The understanding of the processes that lead people to become tourists is of potential interest to anthropologists just as it is to sociologists and psychologists, whose perspectives were discussed in earlier chapters. Taking into consideration the tourism generating area, or the area where tourists come from, using Nash's (ibid.) terminology, tourism is now placed in a wider 'superstructure', linking together the social and economic conditions of generating societies with the effects upon the cultures of destinations. This has led to different propositions and theories of the motivation of tourists that are explained in the next section of the chapter.

The bricoleur, myths and signs

For MacCannell (1976, 1989) tourism can be interpreted as a reaction to the perceived inauthenticity of modern societies with a subsequent search for the authenticity of pre-industrial societies. This idea is similar to the Romantic radical Jean-Jacques Rousseau of a 'noble savage', free from the corrupting constraints of 'civilisation' and benefiting from a 'natural' life. Emphasising the ideas of the symbolic, for MacCannell (1976, 1989), tourism represents a desire to recover the sense of wholeness and structure absent from everyday contemporary life in the urban areas of the West. Tresidder (1999) suggests that this sense of wholeness is often presented in mythological constructions of heritage, wilderness and rurality, which are presented as the antithesis of post-industrial society. He supports this observation through empirical research based upon the content analysis of 18 Irish tourism publications, his analysis of the varying types of landscapes that are presented being described in Box 6.1.

Selwyn (1996) also emphasises the idea that the tourist is someone who 'chases myths'. In this context myths are viewed as a means for resolving intellectual and emotional disharmony that is a means of providing a sense of stability and reasoning to our lives. So in the context of MacCannell's (1976) idea of a search for authenticity, the tourist's quest may for example be structured by a desire to find a society with harmonious social relations, or to discover what they imagine a 'functioning' community should be like.

Think point

Choose a country that you have been to, obtain brochures about it and visit the website of its national tourism board. Analysing the photographic images and descriptions of the promotional material are they similar to your experiences of it? Is there any evidence of a mythological image being created that emphasises a simpler pre-industrial society as in Tresidder's study of Ireland?

Box 6.1 Analysing landscapes of Ireland: chasing the myth?

Tourism can be interpreted as a search for authenticity and the chasing of the 'myth' of a more primitive and simpler life that existed before industrialisation. The tourism industry may subsequently facilitate this myth. Through empirical research, involving the content analysis of 18 Irish tourism publications, Tresidder (1999) identified the following categories of environments and presentation of life in Ireland:

1   *Townscapes*: images of towns typically consist of deserted townscapes with colourful houses or shops. People that are in the images emphasise rural links, being dressed in agricultural wear, driving a horse and cart, or riding a bicycle.
2   *Landscapes*: these are usually deserted, emphasising the wilderness of rural Ireland. If there are people in the landscape, they are isolated couples or a young family.
3   *Heritagescapes*: emphasise a search for our roots and the authenticity of what is lacking in postmodern or post-industrial life.
4   *Workshops*: All scenes of a working rural landscape ignore the mechanisation of agriculture, showing pictures of men and women using pitchforks and other types of manual labour. The emphasis is placed upon displaying a less hurried existence than exists for the majority of people in post-industrial societies.

Source: Tresidder (1999)

Tourism may also be used by individuals to help construct their own identity as social class becomes less important in this role. An often-used term to describe the stage that western societies have reached in their development is post-modernity. Although there exists disagreement over its meaning, its characteristics can be recognised in the social, economic, political and cultural spheres of life (Abercrombie *et al.*, 2000). Particularly, social class is no longer to be held so important in terms of influencing lifestyle and behaviour. This is evident to an extent in tourism, as destinations and types of holidays that were originally the preserve of the upper classes have now transcended many social classes in society.

With a decline in the importance of defining identity and role by social class, the opportunity to create one's own identity in society now exists. Subsequently, personal characteristics, including sexual orientation, have become more important in creating identity than conforming to social norms. In the context of tourism, the tourist becomes a 'cultural bricoleur', using the signs, symbols and artefacts of the different cultures through which they pass creatively to formulate a new identity. Particularly aspects of fashion, for example clothing, hairstyles, body piercing and tattoos, may be taken from other cultures and reassembled into a new identity in the home environment.

The understanding of the use of signs and symbols to convey meaning is associated with the work of C.S. Peirce (1839–1914) in the second half of the nineteenth century. He identified three classifications of symbols: the 'iconic' which represent what they refer to, e.g. a sign showing a ferocious dog; the 'indexical' which can only be interpreted by reference to what is being pointed at in a given context, e.g. the word 'me'

Think point

Are you a 'cultural bricoleur'? How important is social class to your own identity? Have you adopted aspects of clothing or body decoration from other cultures to help establish your own image or identity? If so, where did these influences come from, e.g. friends from different cultural backgrounds, media or tourism?

has obvious different meanings; and the 'symbolic' which are independent of objects to which they refer. For this latter category, meaning is established by convention, the most evident example of this being words in a language.

Tourism is particularly reliant upon iconic interpretation, for example the Eiffel Tower and the Taj Mahal, to signify to tourists that there exists something worth seeing and to lend an identity to place, as Crang (1999: 240) put it: 'making sights out of sight'. The presence of an object such as a famous landmark or some other kind of sign is termed a 'signifier' and the concept that is associated with it 'the signified'. The study of the meaning and the relationship between the 'signifier' and the 'signified' is termed 'semiology' (Burns, 1999).

Signifiers are important, for as MacCannell (1989) points out, the first contact a tourist has with a site is not the site itself but with a representation of it. He extends the concept of a 'signifier' or 'marker' beyond the information attached to a tourist site to include a whole range of different sources of information that first bring it to the tourist's awareness. Included in this potential range of sources could be travel books, accounts of tourists who have already visited the site, travel programmes, cinema and lectures.

The role of using symbols and signs to sell tourism is a further aspect of the semiotics of tourism. In the view of Kinnaird et al. (1994) the tourism industry uses a combination of different signs, symbols, daydreams and fantasies to sell tourism products. For people to voluntarily become tourists implies a belief that what lies away from the home environment is in some way better or different than what they have at home. Subsequently, tourism can be viewed as being concerned with the production and consumption of dreams.

Think point

Which famous 'signifiers' or icons of tourism can you think of, e.g. Eiffel Tower? What messages do they convey? To what extent do you think that the tourism industry is concerned with the production and consumption of dreams?

Authenticity

The concept of the tourism industry selling dreams, and the mythological reconstruction of culture, raises issues about their use as saleable commodities for consumption by tourists. A consequence of this may be a possible commoditising of culture for economic and financial purposes as Boissevain (1996: 11) observes: 'Culture has become a major commodity in the tourism industry.' The use of culture in tourism incorporates many different aspects, including historical monuments, heritage, carnivals and religious ceremonies.

Key issues of the use of culture for tourism include its authenticity, and tourism contributes to cultural change. If authenticity is believed to lie in the past and in more 'primitive' societies then there is an opportunity for the tourism industry to produce or cater for this authenticity (MacCannell, 2001). Thus there exists the paradox that while tourists may be attracted by the authenticity of culture, rituals and other practices may be changed to accommodate the needs and time-frames of the tourism market. Thus a 'staged authenticity' is presented to the tourist; for example, the crocodile ritual performed by the Iatmul people of the Sepik region in New Guinea has been reduced from three days to less than 45 minutes, and in place of its annual performance now takes place upon the arrival of a cruise ship (MacCannell, ibid.). These types of staged events are analogous to what Boorstin (1961) in an early critique of mass tourism referred to as 'pseudo-events'. How tourists will react to the staging of authenticity is unclear. However, whereas in Boorstin's (1961) view, tourists demand superficial experiences, MacCannell (1976) takes the view that tourists demand authenticity.

Yet the ability of tourists to decide what is 'authentic' is debatable. For instance, in the absence of having knowledge that the crocodile ritual originally lasted three days and was performed annually, how could a tourist judge it as being inauthentic on the basis that it had been reduced to 45 minutes? Additionally, in terms of satisfaction with the experience, would the tourist prefer to spend three days of their holiday time observing the ritual or 45 minutes? MacCannell (2001) observes that in terms of tourist satisfaction, a spectacle conforming to pre-given expectations may be more important than authenticity. Thus the spectacle that fulfils the tourist's notion of the authentic may be more satisfying than the real thing. This idea is associated with the theory of hyper-reality developed by Baudrillard (1983), which emphasises that in a consumer culture, images and signs begin to stand in for reality, and that the fake may be so good that the original is no longer required.

The question of authenticity also extends to souvenirs, which as Hitchcock (2000) points out, have different functions. These include the connecting of different social worlds through their production and sales; the purchasing of souvenirs is often one of the few occasions that tourists and local people meet, particularly in the case of enclave or cruise tourism (Hitchcock, ibid.).

In trying to understand the concept of authenticity from the tourist's viewpoint, Selwyn (1996) makes a significant separation between 'knowledge' and 'feeling'. 'Knowledge' implies an understanding of authenticity based upon a scientific rationale and subsequently any event would be judged against technical criteria. By contrast, in the absence of knowledge of what constitutes the authentic, a tourist will rely upon their 'feelings' to decide whether what they are viewing is authentic or not. Hence, in the absence of a detailed knowledge of what actually constitutes the authentic, it is quite possible for a tourist to believe that they are participating in an authentic cultural experience even when they are not.

Think point

What is meant by 'staged authenticity'? From a tourist imperative does it matter if we are presented with staged authenticity in the belief that it is 'real'?

In terms of the impacts of 'staged authenticity' upon the local culture, these are debatable. While the commoditisation of rituals and their performance for money may lead to a loss in their significance, an alternative view is that the interests of tourists in a culture can lead to renewed local pride in its traditions and history. Nor should it be assumed that local communities would be passive receptors of cultural change.

Local communities may also use staged authenticity to protect their local culture. MacCannell (1989) adopts the concept of 'front' and 'back' regions, developed by Goffman (1959), to illustrate how staged authenticity can be used in this way. The 'front' refers to the meeting place of the tourists and community where staged authenticity takes place. The 'back' is the zone to which local people retire, where they conduct patterns

Figure 6.1 Zulus performing a ritual dance at a cultural and heritage centre outside Durban in South Africa. An example of 'staged authenticity' leading to a loss of cultural significance or a new-found cultural pride?

of normal social interaction. Thus 'staged authenticity' can protect the privacy or 'back regions' of local inhabitants by keeping tourists' interests focused upon the commercialised 'front region' (Boissevain, 1996). This theme is developed in the next section of the chapter.

A further consideration of authenticity is who controls culture and how it is presented to the tourist. As Dicks (2003: 58) comments: 'Authenticity is not an objective quality but a subjective judgement, always open to contestation and dissent through conflicting interests.' For example, conflict may arise not only over the use of a culture for tourism but also over how it is presented. Conflict over presentation could potentially arise for example between the interests of tourism businesses and indigenous culture. For instance, tourism businesses may want to promote a type of display aiming to meet the preconceived ideas of the tourists, while an indigenous people may want to present a spectacle that provides an educational message of their culture to a wider audience. These two aims need not be mutually exclusive, but the control of how culture is used and presented will be influential in deciding how the 'authentic' manifests itself.

However, the notion of a tourist purely as a seeker of structure and authenticity based upon a feeling of alienation is debatable (Selwyn, 1996). In an economic system based upon consumption, tourists may enter into the activity of consuming without necessarily feeling a sense of alienation. Franklin (2003), referring to the origins of mass domestic tourism for the proletariat described in Chapter 1, argues this was already the case. Commenting on working-class holidays in the twentieth century, when incomes were for the first time at a level to permit the taking of an annual holiday, he comments (2003: 11): 'their holiday to places such as Blackpool or Brighton, offered the prospect of transition into the consumer world. These places were magical and compelling precisely because they initiated them into the bright and dizzy world of emerging consumerist modernity.' Thus it can also be argued that far from taking the tourist into the past in search of the primitive, in fact tourism may place the tourist into the world of consumerism.

Type of tourists

Although the claim to identify different types of tourists can be made by sociologists, psychologists and anthropologists, a significant typology was developed by Cohen (1979) based upon the experiences of tourists. This typology was developed in part as a reaction to an earlier debate between the two polarised positions taken by Boorstin (1961) and MacCannell (1976) over what a tourist was actually seeking. To reiterate this debate, for Boorstin (1961), mass tourism represented nothing more than a form of mindless escapism, with gullible tourists who would be satisfied with inauthentic 'pseudo events' which were a reflection of their superficial lifestyles at home. In contrast, for MacCannell (1976) tourism is a reaction to the perceived lack of authenticity of modern societies that results in a search for the authentic. Arguing that these polarised positions were too simplistic, Cohen (1979) suggested a typology, shown in Figure 6.2, based upon the experiences of tourists.

Cohen (1979) identifies five different types of typology of experience. A key factor to explain the differences is the degree of psychological and emotional attachment the

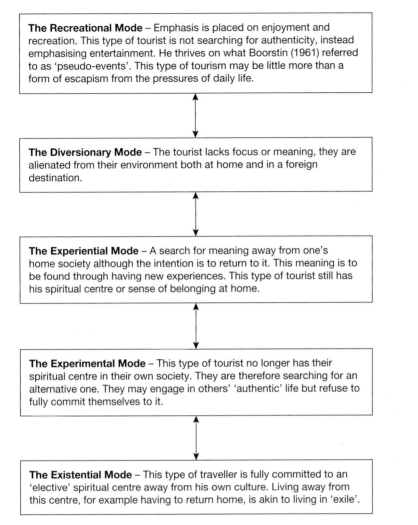

The Recreational Mode – Emphasis is placed on enjoyment and recreation. This type of tourist is not searching for authenticity, instead emphasising entertainment. He thrives on what Boorstin (1961) referred to as 'pseudo-events'. This type of tourism may be little more than a form of escapism from the pressures of daily life.

The Diversionary Mode – The tourist lacks focus or meaning, they are alienated from their environment both at home and in a foreign destination.

The Experiential Mode – A search for meaning away from one's home society although the intention is to return to it. This meaning is to be found through having new experiences. This type of tourist still has his spiritual centre or sense of belonging at home.

The Experimental Mode – This type of tourist no longer has their spiritual centre in their own society. They are therefore searching for an alternative one. They may engage in others' 'authentic' life but refuse to fully commit themselves to it.

The Existential Mode – This type of traveller is fully committed to an 'elective' spiritual centre away from his own culture. Living away from this centre, for example having to return home, is akin to living in 'exile'.

Figure 6.2 Cohen's typology of tourist experiences (after Cohen, 1979).

individual has to their home environment. This subsequently influences the level of intensity and depth of meaning of the experience they desire as a tourist. For example, in the 'recreational mode' an emphasis is placed upon 're-creation' through enjoyment, in preparation for the return home. Thus the destination is merely a place to visit to serve a purpose, rather than there being any spiritual attachment or having a sense of belonging to it. In contrast at the other extreme, the 'existential mode' suggests that in fact the individual could no longer be regarded as a tourist as the destination has become their spiritual base, the place that they feel they emotionally belong to. The life here provides the authenticity of meaning and purpose that they feel is lacking at home. By not being able to relocate there, typically because of employment or family obligations, they are in essence living in a form of 'exile'. Similar to most typologies, the categories are not fixed, and any one individual may or may not search for these different types of experience from tourism during their lifetime. However, the typologies do suggest that the experiences and meanings tourism can provide to individuals are more complex than simply a search for 'escapism' or 'authenticity'.

Think point

Referring to Figure 6.2, how do these types of experience relate to the one's you have had as a tourist?

Tourism as a ritual and sacred journey

A distinctive anthropological perspective of tourism is the concept that the tourist is undertaking a ritual or sacred journey. In Chapter 1 the historical association between religious pilgrimage and tourism was described, and in secular societies there is an argument that tourism itself may be considered as a sacred journey with an aim of spiritual fulfilment. Certainly tourism and pilgrimage display some of the same specific characteristics, for example: the requirement for leisure time; financial resources; and both are condoned by society (Smith, 1989).

For Graburn (2001), tourism is best understood as a ritual, 'one in which the special occasions of leisure and travel stand in opposition to everyday life at home and work' (2001: 42). In this view tourism takes place as a secular or non-religious ritual, embracing goals or activities that have replaced the religious or supernatural experiences of more traditional societies. These rituals are important as they help us to identify and define what is 'ordinary'. Drawing parallels between tourism and the pilgrimages of earlier societies Graburn (2001: 43) comments: 'Vacations involving travel (i.e. tourism) are the modern equivalent for secular societies to the annual and lifelong sequences of festivals and pilgrimages found in more traditional, God-fearing societies.' However, a key difference remains in the personal belief attached to each activity. That is, religious pilgrimage represents a powerful and serious pursuit of spiritual fulfilment, whereas tourism may be described as more of a search for hedonism and superficial wish fulfilment (Pfaffenberger, 1983).

In the view of Graburn (1989, 2001), two distinct spheres of human life are discernable, one being the sacred/non-ordinary/touristic and the other the profane/workaday/at home. Thus, going away on holiday marks the end of the mundane for a limited period, as do other cyclical 'special' events, such as Christmas, Diwali, Eid-ul-Fitr or birthdays. Graburn suggests that the 'magic' of tourism comes from the process of movement to a non-ordinary setting, emphasising the role of travel as an essential component of the experience of tourism.

Although it can be argued that the physical act of travel has been minimised in terms of time by the advent of the jet age, the process of travel has always been central to the experience of tourism. Historically, the difficulty and the enduring of the 'travail' of the journey of pilgrimage was as much of the overall experience as arrival at the shrine. The role of travel as part of the overall experience of tourism in contemporary society is unclear and an under-researched area. For certain types of holiday it may represent the main focus, e.g. cruise-ship holidays or train-based tours of countries. It could be argued that when travel represents a 'non-ordinary' experience, perhaps being for example the only time of the year a flight is taken, then it may be anticipated and experienced with pleasure. However, when travel is undertaken as part of the 'ordinary' or one's everyday life, it is unlikely it will be anticipated in the same way as for the 'non-ordinary'. Also, given the congestion often associated with transport systems as a consequence of exceeding capacity during peak holiday periods, and the delays associated with increased security and border formalities at many airports post-11 September 2001, the extent to which travel is viewed as a form of 'travail' or anticipated with pleasure is uncertain.

Although Graburn's (ibid.) idea of associating tourism with 'sacred' time and places may possibly seem strange, given that the term is usually associated with religious belief, Tresidder (1999: 141) points out that: 'what becomes labelled and adopted as sacred by society does not have to pertain to religion'. Instead the sacred can be viewed as sites or events that are representative of strong emotions and strong beliefs, representing a type of 'social marker', being viewed as essential to continuity and identity. Thus, landscapes can take on the properties of the sacred, representing something that is rooted and unchanging. The importance of such places for our individual and collective psyches is that they make us feel secure, because we know that we have a refuge to escape from the pressures of modern-day urban life and rapid change (Tresidder, ibid.). This type of therapy is not reliant solely upon the actual physical consumption of these landscapes *in situ*, but also through the cognitive consumption of images in photographs and paintings, and by the fact of knowing that they exist.

Think point

To what extent does tourism represent the 'non-ordinary' for you? Do you feel a sense of spirituality as a tourist or are holidays more a time for hedonism and pleasure?

While tourism may represent a cyclical event of the non-ordinary, the role of tourism in one's life may have consequences for an individual's development that moves beyond the cyclical. Referring to the concept of the 'rites of passage', credited to the French folklorist Arnold Van Gennep, Graburn suggests that our lives are marked by a series of changes in our status. In many societies, rites of passage are marked by births, graduations, marriages and funerals, events that are more significant than those that compose the cyclical non-ordinary. Commenting on the 'relatively individualistic, informal lives of the contemporary Euro-Americans', Graburn (2001: 44) argues that tourism can, in certain circumstances, represent self-imposed rites of passage. For example, this could include people who use tourism, perhaps for a longer time of passage than the typical two-week holiday, to recover from major emotional and life-changing events such as a divorce or redundancy, or participate in types of tourism which impose severe physical or mental tests. There is therefore an implicit desire for reflection, recovery and re-creation before moving on to the next stage of one's life.

For Turner and Turner (1978) the ritual process involves three key stages, the first being the 'separation' stage from the ordinary or routine of everyday life. The second stage is entry into a state of 'liminality' in which the structure and order of normal life dissolves, everyday obligations cease to exist, and new forms of relationship are founded based upon a levelling of structures. The final stage is a state of 'communitas', which Franklin (2003: 49) emphasises as: 'a unique social bond between strangers who happen to have in common the fact that they are in some way travelling or "on holiday" together'. The concept originates from anthropological research into African rituals and Christian pilgrimages, which demonstrate that the normal structures of social differentiation, e.g. social class, age and gender become looser or disappear when in a foreign environment (Graburn, 2001). In this sense the experience of travelling with a group of colleagues, friends or strangers and being in a foreign place can act as a great leveller, bringing the special feeling and close bonding of communitas.

This concept emphasises the group nature that is often a feature of tourism, whether defined closely by friendship or family, or more loosely by nationality, special interest or some other type of social characteristic. In terms of how this liminal state of communitas may manifest itself, Graburn (2001) gives the example of Club Mediterranean clients all dressing in the same fashion in their swimming gear, while Franklin (2003) observes the 'bon ami' of backpackers and their code of friendliness and cooperation on the road. This state of communitas may contrast markedly with the more structured and socially exclusive conditions of the urban areas from where many tourists emanate. It is within this spirit of communitas that holidays often become associated with making friends and finding romance.

Think point

To what extent have you experienced a state of 'communitas' while on holiday? In your view can tourism be interpreted as a 'liminal zone' in which the structures and routines of normal life are no longer present?

However, the view of tourism as a form of sacred journey does not have universal acceptance. For Nash (1996) the focus on tourism as a type of ritual and being sacred overemphasises the essential function of the use of tourism for 're-creation', for a return to the day-to-day living of home. It also overemphasises the likely gratification of tourism at the expense of the social influences that shape it in the home environment. According to Nash (ibid.), the existence of a human need associating tourism with ritual has not been adequately demonstrated. He comments (1996: 45): 'Why not, then, put it aside or in brackets and get on with what can be demonstrated, namely the variety of tourist experiences and reactions, their social causes and consequences?' The next section subsequently considers the zone of interaction where many tourist experiences and reactions are observable.

The tourism destination

A major focus of anthropological research has been upon cultural change attributable to tourism in destinations. Given the geographical focus of anthropological research it is unsurprising that the anthropology of tourism focuses on the effects on cultures in less-developed countries. Two main themes of enquiry are evident, the first being the political anthropology of tourism, emphasising the economic, political and social relationships that exist between the areas where tourists come from and those they visit. The second emphasises the consequential cultural change induced in the communities of destinations by tourism.

The political anthropology of the tourism system considers the causal effects that generate tourism in developed countries, besides the impacts of tourism upon the cultures of peoples in destinations. Thus in this approach, the social and economic conditions that gave rise to mass tourism as explained in Chapter 1 have to be considered, not least the necessity for a level of economic production to provide the resources for leisure. As Nash (1989: 39) comments: 'If productivity is the key to tourism, then any analysis of tourist development without reference to productive centres that generate tourists' needs and tourists is bound to be incomplete.' As we have seen in Chapter 1, the typical characteristics of centres that generate international tourists are that they are urban and have been historically located in the developed economies of western countries, and more contemporarily in the rapidly growing economies of Asia. The historical dominance of the West, in its generation and control of the tourism industry, has led to tourism being likened to a form of imperialism.

Tourism as imperialism

The basis of the concept of tourism as a form of imperialism is that it represents an expansion of a nation's economic and political interests to other countries. Predominantly, the flow of this political and economic expansion is from western countries to less-developed ones. In the context of the political relationship of tourism between the 'developed' and the 'developing' worlds, this can be viewed as a relationship between the 'dominant' and 'subordinate'. Turner and Ash cynically express this relationship in the terminology of 'hosts' and 'guests', with the host community held to be subordinate, at the mercy of dominant guests. In the polemic words of Turner and Ash (1975: 129) the 'guests' represent: 'a form of cultural imperialism, an unending pursuit of fun, sun and sex by the golden hordes of pleasure seekers who are damaging local cultures and polluting the world in their quest'.

Besides generating the conditions that permit people to participate in recreational and leisured tourism, metropolitan centres also have economic and political influences via trade and political channels, as explained in the last chapter. Nash (1989: 39) observes that: 'metropolitan centres have varying degrees of control over the nature of tourism and its development. . . . It is this power over touristic and related developments abroad that makes a metropolitan centre imperialistic and tourism a form of imperialism.'

However, as he points out, the securing of power in other countries is not achieved by military intervention but by a mix of foreign interests often working with a limited local elite. The extension of this power is highly influential in deciding which areas of the world will develop as international tourism markets. Consequently, potential destinations must display the cultural and natural characteristics that correspond to the wishes and desires of the peoples of the metropoles which, with the passing of time, begin to reflect the characteristics of the tourists' home environment. As Nash (1989: 42) comments:

> One cannot begin to account for the character of the Costa del Sol without reference to north-western Europe; turn-of-the century Nice without reference to England and Tsarist Russia; or Miami Beach and Catskills without reference to New York City.

The movement of tourists between developed and less-developed countries consequently brings different cultures into contact with each other. The power relationship of this contact and its consequent cultural impact is of particular interest to anthropologists. However, before proceeding to discuss cultural changes it is necessary to remember that tourism is based upon reciprocity between home and away. Consequently, the experiences that people have as tourists may induce change in the cultures they come from besides those they visit. Also, although the emphasis of anthropological research is upon cultural change induced by tourism in the less-developed world, it is important to remember that most international tourism takes place inside Europe, between cultures that are not economically subordinate.

It is also necessary to remember that power relationships within the tourism system exist not only between countries but also between different groups and classes in society. These will subsequently influence how culture is used for, and affected by, tourism, as is dramatically exemplified in Box 6.2.

A further example of how power relationships influence which aspects of culture are presented to tourists is the decision by the Afghan government to promote Osama Bin Laden's Tora Bora mountain hideout as a tourist attraction, described in Box 6.3.

Box 6.2 Who controls culture?

Commenting on the wishes of indigenous people in Malaysia not to be part of a tourism development plan proposed by central government, Malaysia's Deputy Minister of Tourism, K.C. Chan comments:

> Do they (tribal peoples) want to sit in their longhouses for ever or join a more advanced society? They are so used to their life in the jungle. If they can earn a better living from tourism, why not? It's part of the modernisation process, the 2020 vision. We should not be proud of backward people.

The above quotation highlights the tensions of the relationships that may exist between tourism, identity, economic development and politics. From the inference in the quote, it would imply that tourism was going to be forced by the government upon a people who did not desire it, for the purposes of national development and modernisation.

Source: Vidal (1994)

Box 6.3 'Bin Laden's hideout is touted as a tourist site'

This was the headline from a newspaper article describing how the Afghan government plans to develop the country's fledgling tourism industry. They believe that the caves that were a refuge for Bin Laden and his fighters, together with the presence of old Russian tanks and crashed helicopter gunships from the battles of the Soviet occupation in the 1980s, will act as a 'tourism magnet'. The caves include army barracks, living quarters and tunnels large enough to hide armoured vehicles.

The government is searching for foreign investment to develop the site with three Japanese investors having already visited. The target market for the 'attraction' is the 'adventure tourist'. Further planned developments by the Ministry of Tourism include battlefields of the Soviet occupation.

Source: Coghlan (2004)

Impacts on culture

Given that tourism brings people from different cultures together it is inevitable that they should have an influence upon each other. In the expectation that this would bring a greater understanding between cultures, the United Nations Environment Programme (UNEP) stressed the bringing together of people as a positive aspect of tourism (Lea, 1988). While this optimistic view of tourism is encouraging, Robinson (1999) observes that such benefits are usually localised, being heavily dependent upon the local community not being significantly economically disadvantaged. In contrast to the UNEP's positive vision of cultural enhancement, much of the empirical work that has been carried out in anthropology offers an unfavourable view of the cultural impacts of tourism, with the exception of a few earlier studies that viewed tourism as a benign agent of change.

While it is simplistic to generalise whether the impact of tourism is 'good' or 'bad', there is little dispute that it can induce cultural change. A key anthropological concept to help explain how tourism affects cultures is 'acculturation', which Burns (1999: 104) defines as: 'the process by which a borrowing of one or some elements of culture takes place as a result of a contact at any destination between two different societies.' From a western perspective, cultures in less-developed countries are often viewed as being 'pristine' or existing in isolation of outside influences before tourism. However, in the view of Dicks (2003) the existence of original pristine cultures is unlikely, given the economic interdependence of global trade and subsequent outside influences.

Nor, when assessing the cultural impacts of tourism, is it always easy to separate out those attributable to tourism from other sources such as the global media and information technology. This point is poignantly illustrated in this quote from Giddens (1999: 6):

A friend of mine studies village life in central Africa. A few years ago, she paid her first visit to a remote area where she was to carry out her fieldwork. She expected to find out about the traditional pastimes of this isolated community. Instead, the occasion turned out to be a viewing of Basic Instinct on video. The film at that point hadn't even reached the cinemas in London.

In the view of Robinson (1999) tourists can influence cultural change in at least three ways, as is suggested in Figure 6.3.

The notion of the tourist as a 'culture prophet' places an emphasis upon the demonstration effect of the tourist inducing cultural change to a greater extent than other types of influence such as media. Tourists may also act as a 'catalyst' to change, speeding up cultural changes that had already started in response to other influences, e.g. from information technology or trade. It is also important to consider that there are few communities left in the world from which at least a few members have not visited other cultures. Subsequently, the participation of local people in tourism will also be likely to induce cultural change.

The last type of change identified by Robinson (ibid.) is where the tourist acts as an inhibitor of cultural change, encouraging societies to conserve and protect their own culture. The likelihood of this occurring is perhaps less common than in the latter two cases. However, the feedback from tourists to local people about aspects of their culture that they like, may engender a sense of pride that had been absent or dwindling. Subsequently, people may take a more positive view of their own culture, encouraging them to conserve it or stop it being diluted by outside influences (Boissevain, 1996). Nevertheless, in cultures that possess a material standard of living that is greatly exceeded by the tourist, the behaviour of people on holiday may be a very persuasive one. This is particularly so if there is no conception of the economic and social stresses, e.g. the need to work long hours, crime and family breakdowns, that characterise western societies.

Sometimes, a consensus of local opinion may exist that tourists demonstrate values that are in conflict with their own, hence making tourist behaviour unattractive to local people. The demonstration effect of tourists may conflict with religious beliefs, e.g.

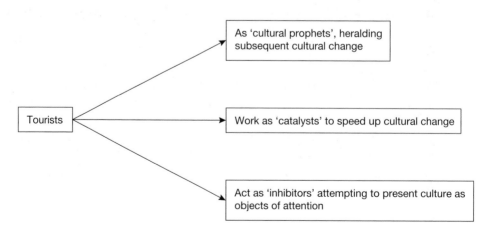

Figure 6.3 The influences of tourists on cultural change (after Robinson, 1999).

tourists dressed inappropriately in swimwear to visit mosques. Also, given that tourism offers a potential escape route from established routines, this may manifest itself in inappropriate behaviour, as Boissevain comments (1996: 5): 'Strange dress and weakening inhabitations are not infrequently accompanied by behaviour that would be quite unacceptable at home. It can be loud, lecherous, drunken and rude. In short, many tourists, for various reasons, are occasionally most unpleasant guests.' Local people may subsequently respond by instructing tourists how to behave, as is shown in Figure 6.4. This photograph taken in the Annapurna area of Nepal is of a notice board produced to persuade tourists not to give local children sweets, pens or money, in a bid to arrest a culture of begging that has developed since the arrival of western tourists to the area.

The possible stages of cultural reaction to tourism that may occur in a community is summarised in Figure 6.5, based upon Doxey's Irridex (1975).

However, although Doxey's original model may have a simple appeal, it does not have the empirical base to support it as a theory (Burns, 1999). It is also pre-deterministic and it cannot be assumed that a community will necessarily pass through all the stages of the model. It is also important to guard against a generalisation of 'community' reaction to tourism. The concept of community is problematic; few communities are homogenous, usually encompassing the dynamism of a wide range of groups of varying political and social views. It would be reasonable to expect that those who financially and economically benefit from tourism are likely to be more favourable towards it, while more marginalised groups may possibly view tourism less favourably. This model is additionally complicated by the differences of the cultures that tourists are coming into contact with and also by the type of tourism and tourists in a destination.

Figure 6.4 Save our self-esteem.

Euphoria – Local people are pleased to see tourists, as they are a novelty. Numbers are low and the facilities for tourism, such as they are, are owned by local people. The type of contact taking place is only slightly commercial.

Segregation – Residents begin to separate their lives from those of tourists as the numbers of tourists begin to increase. Tourism also becomes more commercialised as more entrepreneurs see the financial potential for tourism. The nature of the relationship becomes increasingly centred upon business.

Opposition – More tourists continue to arrive in a highly commercialised arrangement with foreign tour operators. The destination develops with outside investment. Some groups in the local population become opposed to tourism because of a restricted access to resources and unacceptable social and cultural behaviour of tourists. The type of tourists changes, there is a confrontation between local people and tourists.

Antagonism – Some groups in the community may openly protest against tourism. This may manifest itself in direct attacks on tourists or tourist facilities.

Figure 6.5 Stages of possible cultural reaction to tourism (after Doxey, 1975).

Think point

Have you observed any influences of tourism upon cultures you have visited? Referring to Figure 6.5 how have local people interacted with you as a tourist? Have you noticed differences between various destinations? If so, what factors do you think account for this?

Nevertheless, as the tourist is paying for the privilege of being free from the constraints of everyday life and requires a servile class to service their needs, tourism inevitably involves a degree of segregation. MacCannell (2001) points out that the common denominator of all tourism is that it takes place away from home, and constitutes a break from the daily distraction and responsibility of home and work. So while the tourist is engaged in leisure, the host is normally engaged in work (Krippendorf, 1986). This relationship of the served and the server automatically implants social barriers to the development of a meaningful relationship, both on grounds of culture and social class (Nash, 1989).

The potential for opposition and antagonism to tourism from local people is likely not just to be a function of numbers but also a consequence of tourists' knowledge of and respect for local traditions and customs. For example, Smith (1989) describes the case of the younger Inuit who resented the behaviour of early foreign tourists coming to their home in northern Alaska. Smith recounts how these tourists roamed the beaches, photographing subsistence activities such as seal butchering and sneered at the odour of drying fish, the riposte from the local youth being graffiti saying 'Tourists, Go Home'. Another reaction was for fishermen to erect screens to shield their work and eventually hire taxis to take the seals home so they could continue their work in private (Smith, ibid.). The theme of communities protecting their cultures is expanded upon in the next section of this chapter.

Protecting culture

As was suggested by Robinson (1999), one possible reaction from a community is to seek to protect their culture from tourism while living alongside it, by creating a 'back region' protected from tourism. The protecting of culture is a strategy employed by a number of local communities, utilising a variety of methods, identified by Boissevain (1996) based upon empirical fieldwork and summarised in Figure 6.6.

Covert resistance is characterised by the mundane and daily struggle of the weak against the powerful, but avoids direct defiance. According to Boissevain (1996: 14): 'Examples of this covert, low-key resistance are the sulking, grumbling, obstruction, gossip, ridicule, and surreptitious insults directed by the weak at the more powerful.' For example, given that employees in the tourism industry rely upon the goodwill of tourists they are unlikely to confront them directly, but display attitudes including sullenness, haughtiness and rudeness. Other aspects of this covert resistance include grumbling, gossiping and the stereotyping of difficult tourists. The outcomes of such covert resistance are that they enable persons subordinated by tourism to retain their self-respect.

Communities who do not wish to display all elements of their culture to tourism may deliberately *hide* parts of it from tourists. This may include for example keeping certain

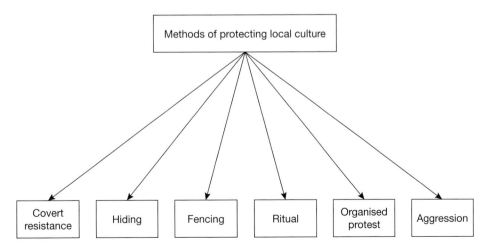

Figure 6.6 Methods of protecting local culture (after Boissevain, 1996).

foods for themselves. Notably, celebrations and festivals may be deliberately held at times which guard them from the attention of tourists, thus becoming 'insider-only' celebrations. The strategy of *fencing* can be interpreted both metaphorically and literally, emphasising the presentation of space that is free from tourism, in what is often 'contested territory' between the needs and wants of residents and tourists. This can involve the physical fencing off of space from tourists, or the relocation of activities to a new spatial area free from the tourist gaze.

The use of *ritual* in a community helps to combat the stress of uncertainty and change, which tourism can bring. Ritual has historically been an important means of challenging the threat of externally exposed change, such as epidemics of illness or foreign invasion. Tourism may encourage the resurgence of rituals as a means of confronting any perception of destructive change, becoming a means of re-establishing camaraderie, identity and values. Although to date fairly rare, the impacts of tourism may result in *organised protests by local people* against tourism. Typically this form of protection is a result of confrontation over resources or where tourists are challenging indigenous values and morals. *Aggression* is the extreme end of the reaction to the impacts of tourism. Citing the work of Van den Bergh (1994), Boissevain (1996) gives the example of a French tourist who was stoned to death by the villagers of San Juan Chamula in Chiapas in Mexico for photographing their carnival. Tourists may also be deliberate targets of crime, such as mugging and handbag snatching.

Summary

- Anthropology is concerned with understanding the culture of other societies, and its origins are associated with the time of European colonial expansion during the nineteenth century. The expansion of mass international tourism in the latter half of the twentieth century meant that tourists were visiting countries in which many anthropologists had carried out their fieldwork, and cultures were interacting on a scale never seen before. A defining characteristic of anthropology is its methodology of ethnography, which permits the understanding of nuances of tourism that may be missed by other research methodologies.

- The principal actor of cultural change is the tourist. From an anthropological perspective, an important reason for people becoming tourists is a search for the 'myth' of authenticity, which has disappeared in developed countries since the onset of the Industrial Revolution. Tourism may be used as a means of constructing one's own identity in society as social class becomes less important as a determinatory factor. Thus the tourist may become a 'cultural bricoleur', using the signs, symbols and artefacts of the different cultures through which they pass to formulate a new identity.

- The possibility of the tourist industry mythologically reconstructing culture raises the issue of authenticity and the use of culture as a type of commodity. Consequently, a culture may become commoditised for economic and financial purposes leading to a type of 'staged authenticity'. Yet, from the point of view of the tourist, a spectacle based on the 'hyper-real' that meets their expectations may be more satisfying than the real thing. In terms of its impact upon the local culture, it can be argued that the commoditisation of rituals may lead to a loss in their significance, or alternatively lead to a renewed sense of pride and interest in local traditions. Staged authenticity may also protect the 'back regions' of local inhabitants by keeping tourists focused on the commercialised 'front region'. The power relationships of who determines the 'what and how' of culture is presented as critical in determining its outcomes.

■ A distinctive anthropological perspective of tourism is the concept that the tourist is undertaking a ritual or sacred journey. Tourism may be viewed as a secular ritual, embracing goals or activities that have replaced the religious experiences of traditional societies. It can be seen to represent a time of the 'non-ordinary' and a type of cyclical special event, similar to birthdays or religious festivals such as Eid-ul-Fitr, Diwali or Christmas. Tourism may also have consequences for individual development that extend beyond the cyclical, acting as 'rites of passage'. However, the view of tourism as a sacred journey does not have universal acceptance, being seen to overemphasise the likely gratification of tourism, at the expense of the social influences that shape it in the home environment.

■ The political anthropology of tourism considers the causal effects that generate tourism in developed countries besides its impact upon culture. The dominance of the West in the generation and control of tourism has led to it being likened to a form of imperialism. Tourism may also be viewed as a type of superstructure, having effects on the societies tourists return to besides the ones they visit.

■ Tourism can induce cultural change, which is more likely when communities are economically disadvantaged compared to the tourist. A key anthropological concept to explain how tourism affects culture is 'acculturation'. However, the presumption that tourists will always influence culture in an involuntary fashion is too simplistic. In place of taking the tourist as a role model, communities may deliberately take steps to protect their culture.

Suggested reading

Boissevain, J. (ed.) (1996) *Coping with Tourists: European Reactions to Mass Tourism*, Berghahn Books, Oxford.

Burns, P. (1999) *An Introduction to Tourism and Anthropology*, Routledge, London.

Franklin, A. (2003) *Tourism: An Introduction*, Sage Publications, London.

MacCannell, D. (1989) *The Tourist: A New Theory of the Leisure Class*, 2nd edn, Macmillan Press, London.

Nash, D. (1996) *Anthropology of Tourism*, Pergamon, Oxford.

Selwyn, T. (ed.) (1996) *The Tourist Image: Myths and Myth Making in Tourism*, John Wiley & Sons, Chichester.

Smith, V.L. and Brent, M. (eds) (2001) *Hosts and Guests Revisited: Tourism Issues of the 21st Century*, Cognizant Communications Corporation, New York.

ENVIRONMENTAL STUDIES AND TOURISM

7

This chapter will:

- introduce environmental studies;
- outline the historical interaction between tourism and the environment;
- consider how tourism impacts negatively upon nature;
- describe how a destination's natural environment may change with its life cycle;
- critically discuss tourism's role in environmental conservation;
- explore the development of national parks;
- consider how stakeholders' environmental ethic influences the relationship between tourism and the environment.

Introduction

The previous chapter was based upon issues of the relationship tourism has with culture. This chapter concentrates upon the relationship tourism has with the physical environment. This relationship can be viewed as being especially important, given that the resources and attractions of nature are the *raison d'être* of much of tourism. Tourism's interaction with the environment is complex: it can help conserve resources, have negative impacts, and can itself be threatened by human-induced changes in the environment. Using a definition of the environment as the non-human surroundings of the world that we inhabit, this chapter subsequently discusses the positive and negative interaction of tourism with nature, drawing upon some of the key concepts that have emerged within the multi-disciplinary area of environmental studies to enhance our understanding of this relationship.

Environmental studies and concerns

The emergence of environmental studies as a subject for academic study can be dated to the 1960s. Its development was a consequence of the realisation that human activity was impacting upon and creating a state of disequilibrium in the environment. Instrumental to this understanding was scientific enquiry that has made us aware of many of the environmental concerns with which we are familiar today, e.g. global warming, ozone depletion, biodiversity loss and pollution. However, although science has highlighted these problems it does not necessarily have the solutions to them. There also exists a high level of scientific uncertainty over the magnitude and timescale of many of the environmental effects of human-induced change.

Given that many environmental problems are seemingly a consequence of human behaviour, the social sciences have an essential role in understanding and helping to find solutions to them. Hence, a suitable definition of environmental studies would be the one given by Nelissen *et al.* (1997: 13) as: 'We define environmental studies as the interdisciplinary field of studies concerned with problems in the relationship between man, society, and environment.' Although not specified by the authors, it is assumed that in this definition the term 'man' is used in a neutral sense, referring to both men and women. Nelissen *et al.* (ibid.) comment how the 'domain' of environmental studies, which they also choose to refer to as a 'discipline', has matured considerably from the late 1980s with most universities in the western world now offering courses in environmental studies. This is perhaps unsurprising when the changes in the environment in the second half of the twentieth century are considered, as shown in Box 7.1.

This list is far from comprehensive and there are critics who would choose to dispute this scientific evidence. However, besides the support of scientific measurement it is perhaps our own observations of environmental changes that are persuading many of us of the human impact upon nature. According to Hodgson (1996) the first incident that created public concern about the state of the environment was the poisoning of water and fish from chemical discharges from the Chisso Corporation factory in Minamata Bay in Japan in the mid-1950s, leading to subsequent damage to people's nervous systems and a high incidence of birth defects and brain damage. Yet, as Hodgson (ibid.) points out, during the 1950s many people in the world also regarded smoke and dirt as signs of industrial progress, especially in a post-Second World War world that was attempting to reconstruct national economies. However, a decisive moment in the questioning of scientific progress came in 1962 with the publication of biologist Rachel Carson's book *Silent Spring*, which heavily criticised the ecological damage resulting from the use of agro-chemicals on farmland in the US and emphasised the dangers to human health that could be passed through the food chain. The book became a bestseller and had a

---

Box 7.1 Human-induced changes in the environment

- Atmospheric CO_2 levels are 30 per cent higher than in pre-industrial times.
- The 1990s was the warmest decade since written records began.
- Twenty-seven per cent of the world's coral reefs are threatened.
- Fifty per cent of the rainforests have been destroyed post-1950.
- Twenty-four per cent of the total species of mammals are threatened with extinction.
- The illegal trade in wildlife is the second largest illegal trade in the world after drugs.

major influence on public consciousness and subsequently on regulatory policy, with the banning or restriction of use being placed on twelve of the pesticides and herbicides that Carson identified as being most dangerous, including the notorious DDT.

As western societies became progressively dependent upon oil during the 1960s to power factories, heat homes and to fuel the increasing number of vehicles, in 1967 the first major 'environmental disaster' in the West occurred. In March of that year, the Torrey Canyon oil tanker carrying a full load of 120,000 tonnes of crude oil hit rocks off the coast of England, leading to the release of oil onto its southwest coast. This was the first time anything like this had ever happened, as the oil polluted the water and washed onto the beaches. Media images of birds with black oil and tar stuck in their feathers stopping them flying were poignant, as was the knowledge that the oil had killed fish and shrimps and other forms of life. Besides causing a high level of public concern, the breakup highlighted the fact that a higher material standard of living did not come free of environmental risk and costs. In the same year, the first major oil spill in the US occurred from an off-shore rig near Santa Barbara, releasing millions of tonnes of crude oil onto the coasts of California. In 1969, toxins leaked into the River Rhine, poisoning millions of fish and threatening the quality of drinking water for millions of Europeans (Dalton, 1993).

Subsequently, there was a growing body of evidence that industrial growth and progress did not come free of environmental cost. In 1968, perceptions of the world as having unlimited and abundant resources were also challenged by the first widely broad-cast television images of the earth shot from the American spacecraft *Apollo 8*, showing the earth as a sphere floating in space. The concept of a 'spaceship earth' was the subject of a famous essay in environmental studies by Boulding (1973), questioning the 'cowboy economy' associated with the reckless and exploitative use of nature, which he believed typified the western approach to development. In place he argued that we should begin to conceptualise the earth as having a 'spaceman economy', in which the earth like a spaceship doesn't have unlimited reserves of anything, and in which humans must find their place without threatening its cyclical ecological system.

New-found public concern over the environment manifested itself in 'Earth Day', which took place in the US on 22 April 1970. Organised by Senator Gaylord Nelson, an estimated 20 million Americans participated, with 100,000 converging on New York's Fifth Avenue. As Spowers (2002) observes, this large demonstration pressured American politicians into being seen to do something on the environment. In 1972 the first international conference on the environment, organised by the United Nations, was held in Stockholm, one outcome being the United Nations Environment Programme (UNEP), which is influential in directing global environmental policy today. Evidence of an increasing public environmental concern was also evidenced by the founding of major environmental NGOs including Greenpeace and Friends of the Earth in the early 1970s. One particular focus of Greenpeace was to campaign against whale hunting, as many species of whale faced extinction from over-hunting (Hodgson, 1996). The role of NGOs was important in attracting and maintaining media attention on the environment and lobbying government about environmental issues.

In 1972 the *Limits to Growth: A Report for the Club of Rome Project* was also published. This report was the result of collaborative research between a group of scientists and business leaders into population growth, resource use and other environmental trends. Predictions of pollution, resource depletion and heightened death rates resulting from a lack of food and health services drew attention to environmental issues even if they have been proven incorrect with the passage of time. In 1979, the near meltdown of the nuclear reactor at Three Mile Island in Pennsylvania in the US and the subsequent threat of a major environmental and civil catastrophe alerted the public to the dangers of the nuclear power programme being pursued in many counties. Opposition to nuclear power became a central focus of the environmental movement in the 1970s, based upon both the environmental consequences of the programme, and its strong link to the production of plutonium for nuclear weapons.

By the 1980s, environmental problems resulting from human actions had become regular media items. The explosion in 1984 at the Union Carbide factory at Bhopal in India, killing 2,000 people and maiming 20,000 more, raised the issue of corporate responsibility to the environment, the company being forced to pay US$ 470 million compensation (Hodgson, 1996). Even though the plant is no longer used and under the control of the Indian government, problems of pollution of the local water supply, high cancer rates, and ongoing claims for compensation are still issues more than 20 years later.

Global warming associated with the burning of the earth's carbon stocks to supply the energy for an increasing consumer society and the associated release of 'greenhouse gases' (GHG) gained increased media coverage in the 1980s, as did the depletion of the ozone layer. The predicted climatic changes associated with global warming, and the increased risk of melanoma or skin cancer resulting from ozone depletion, became issues of human welfare. Disquiet was also being voiced by NGOs over the rate of depletion of the most biodiverse ecosystem in the world – tropical rainforests – for the purposes of agriculture and logging. Nuclear power remained an ongoing concern as the world experienced its worst nuclear disaster to date with the meltdown of the nuclear reactor at Chernobyl in the Ukraine in 1986, the effects of the nuclear fallout being felt across Europe. One year later, the 'European Year of the Environment' was held in the European Union. In 1988 the World Commission on Environment and Development (WCED) report *Our Common Future* was published, which emphasised a re-evaluation of global resources, and advocated development based upon sustainable principles to arrest degradation of the environment, as was explained in Chapter 5.

During the 1990s, campaigns for the rights of animals and against animal experimentation became more vocal and violent. Protests against road building became a focus for environmental campaigners in Britain and other European countries as concerns grew over the pace of the loss of countryside and nature. Green politics in Europe gained increasing recognition through formal political routes, notably the formation of a governing red–green coalition in Germany, and by the end of the decade green politicians were in charge of the environment ministries of Germany, France, Italy and Finland (Bowcott *et al.*, 1999). Concerns over the practices employed by farmers were also heightened with the outbreak of Bovine Spongiform Encephalopathy (BSE) in Britain, which not only threatened animal life, but could also be transmitted to humans in the form of Creudtzfeld-Jakob Disease (CJD). The economic impacts of this were strongly felt by farmers as British beef exports were boycotted around the world. Worries over genetically modified crops were also raised in Europe and there has been a subsequent increased demand for organically produced vegetables, fruits and meats. The emergence of a new 'green' consumer culture represented another channelling of environmental action (Burchell and Lightfoot, 2001).

The 1990s also marked a time of collective international policy-making on the environment. In 1992 in Rio de Janeiro, more than 140 countries were represented at the United Nations Conference on Environment and Development (UNCED), more commonly termed the 'Earth Summit' referred to in Chapter 5. Alongside this official event, 20,000 members of NGOs from around the world met at a separate event, the 'Global Forum' (Spowers, 2002). Two successive Earth Summits followed this in New York in 1997 and in Johannesburg in 2002. Although the Earth Summits have received criticism for being nothing other than sophistry to satisfy public opinion, an outcome of the summits has been two biodiversity conventions, a commission on sustainable development, and Agenda 21 which sets out guidelines for governments to follow. In 1997, the 'Kyoto Protocol', the world treaty attempting to control global warming, was agreed. The aim of this treaty is to reduce greenhouse emissions, although the US, which produces 25 per cent of the total (McCarthy, 2003), has still to ratify the treaty to lend it a meaningful application.

Think point

What do you think are the major environmental problems facing the planet today? Contrast your awareness of environmental issues with those of your parents when they were your age.

The history of tourism's relationship with the environment

While concerns about the effects of human activity upon the environment had become a focus of scientific enquiry and merited media coverage by the 1960s, tourism received little consideration in this context. If reference was made to tourism's relationship with its natural surroundings it was usually in the context of a 'smokeless industry'. This perception was enhanced by the imagery of tourism, embracing virtues of landscape beauty and virginity, portrayed in photographs of 'exotic' beaches and mountain areas in sunshine. Such images presented a stark contrast to the smoke and dirt that characterised economic progress based upon the heavy industrial production of western urbanised areas during the first half of the twentieth century.

However, there were a few dissenting voices to an acceptance of the 'smokelessness' of tourism. Milne (1988) comments that in the early 1960s there was concern being expressed over the possible ecological imbalance that could result from tourism development in Tahiti in the Pacific. The observation of the effects of increasing numbers of people descending upon beautiful areas of southern Europe in the 1960s led the economist Mishan (1967: 141) to write:

Once serene and lovely towns such as Andorra and Biarritz are smothered with new hotels and the dust and roar of motorised traffic. The isles of Greece have become a sprinkling of lidos in the Aegean Sea. Delphi is ringed with shiny new hotels. In Italy the real estate man is responsible for the atrocities exemplified by the skyscraper approach to Rome seen across the Campagna, while the annual invasion of tourists has transformed once-famous resorts, Rapallo, Capri, Alassio and scores of others, before the last war no less enchanting, into so many vulgar Coney Islands.

During the 1970s, questions about the environmental impacts of tourism began to be raised more widely, as the volume of international tourism increased and its spatial boundaries spread further. Recognition of the problems that could be caused by tourism led the Organisation of Economic Cooperation and Development (OECD) in 1977 to establish a group of experts to examine the interaction between tourism and the environment. Negative environmental impacts of tourism, including a loss of natural landscape, pollution, and the destruction of flora and fauna, were observed. Concerns were also expressed in academic circles, vividly in *The Golden Hordes* by Turner and Ash, their views aptly summated in the following passage:

Tourism is an invasion outwards from the highly developed metropolitan centres into the 'uncivilised' peripheries. It destroys uncomprehendingly and unintentionally, since one cannot impute malice to millions of people or even to thousands of businessmen and entrepreneurs. . . . As a mass movement of peoples tourism deserves to be regarded with suspicion and disquiet, if not outright dread.

(1975: 127)

Similarly polemic in his view was Goldsmith (1974: 10) who comments:

A large part of the coast of Southern Spain, of the South of France and of the Italian Riviera have already been mutilated beyond redemption with countless hotels together with their associated amenities. An island that has suffered particularly from tourism is Hawaii. This once beautiful island is disfigured with countless skyscrapers.

It is therefore apparent that the scale of tourism and its ability to aesthetically change landscapes and impact upon the environment was an active concern by the 1970s.

In the 1980s, the spread of mass tourism beyond the Mediterranean into new areas including South East Asia, Africa and the Caribbean meant that there was increasing focus on tourism as a form of economic development by the United Nations and national governments in less-developed countries, as was discussed in Chapter 5. Besides the economic aspects of development, concerns over the environmental and cultural consequences of tourism development were being recognised by NGOs and parts of the academic community. NGOs such as Tourism Concern, the UK-based campaigning group for humane tourism development, and the Ecotourism Society in the US were established in the late 1980s to promote ethically based tourism for both indigenous peoples and nature. Local pressure-groups, concerned by the effects of tourism development on their physical and cultural environment, were also formed, such as the Goa Foundation in the state of Goa in India. The Goa Foundation has opposed tourism because of the restriction or loss of access to essential resources such as water and electricity and human-rights violations associated with its development.

At the end of the 1990s tourism development was attacked directly by militant environmental campaigners or 'eco-warriors' for the first time. Ski facilities were burnt down in Vail in Colorado at the beginning of 1999 because of their possible impacts upon wildlife. In the 1990s the tourism industry itself began to take action over the environment, with many tour operators, hotels and airlines attempting to improve their environmental credibility. Growing concern over the environmental impacts of tourism also found its way into the popular press, raising levels of public awareness. The need for a more sustainable approach to tourism development also led NGOs and charities not directly associated with tourism, such as the World Wide Fund for Nature, Voluntary Service Overseas, Tear Fund and Oxfam, to become involved. The tourism market also diversified, with a significant market niche demonstrating a desire for 'unspoilt' and 'authentic' environments. A summary of the changing attitudes towards the relationship between the environment and tourism over the last five decades is shown in Table 7.1.

Table 7.1 Changing attitudes to tourism (after Hudman, 1991; Holden, 2000).

Decade	Attitudes to tourism
1950	Enjoy – international tourism still restricted to a relatively small elite; high levels of participation in domestic tourism.
1960	Enjoy – quickening pace of 'mass' participation in international tourism; early expressions of environmental concern over tourism development.
1970	Increasing awareness in academic circles that tourism is not a 'smokeless industry' – mass tourism arrives in the eastern Mediterranean; OECD establish a working committee on tourism and the environment; publication of Turner and Ash's *Golden Hordes*.
1980	By the end of the 1980s tourists began to desert traditional locations such as parts of coastal Spain which were seen as passé and over-developed; tourism increasingly viewed as a development tool for less-developed countries, founding of tourism pressure groups such as Tourism Concern (UK) and the Ecotourism Society (US).
1990	'Eco-warriors' target tourism development in Colorado. More tourists becoming environmentally aware. The tourism industry begins to respond to concerns over the environment. 'Eco-tourism', 'green tourism', and 'sustainable tourism' become popular phrases.

Tourism and the environment – a two-way relationship

Tourism is dependent upon the use of natural resources and will consequently usually have either a negative or positive impact upon nature. This relationship is a reciprocal rather than a linear one, i.e. while tourism may impact upon nature, changes in the environment in which tourism takes place may also influence tourism. Notably, climate change associated with global warming poses a particular threat to the future of tourism as the WTO (2003c: 8) comment: 'In two environments which are vital for tourism activities and where tourism is an equally vital component in regional and local economies – coastal zones and mountain regions – climate change puts tourism at risk.' It is evident from this statement that environmental change can threaten not only the nature of destinations but also the economic and social well-being of their populations.

This section of the chapter subsequently discusses the duality of the tourism and environment relationship, the impacts that tourism can have upon the environment, and the consequences that changes in the environment can have for tourism. It will attempt to be as holistic as possible, although our knowledge of the totality of human impacts upon nature is limited, and subsequently so in the case of tourism. This limitation of our insight is a consequence of a number of factors, including: research into impact studies is a relatively immature subject and a true multidisciplinary approach to investigation has yet to be developed; research into the environmental consequences of tourism tends to be reactive and it is subsequently not always easy to establish a baseline against which to monitor the degree of change; separating out the environmental impacts attributable to tourism from the effects of other economic or social activities may be problematic; similarly, separating the source of impacts upon the environment attributable to local residents or tourists can be difficult; and the consequences of tourism are difficult to monitor and assess because tourism development is often incremental and the effects are cumulative (Holden, 2000; Hunter and Green, 1995; Mieczkowski, 1995). Additionally, spatial discontinuities are inherent to tourism, for example the effects of air pollution caused by air and car emissions may contribute to acid rain that destroys forests hundreds of kilometres away.

Although the environmental impacts of tourism are diverse, two broad categories are discernible, i.e. the negative and the positive. To provide a structure to the discussion, the negative impacts of tourism are explained first, followed by the positive environmental consequences that tourism has for the environment. This ordering is reflective of much of the observation and commentary that has been expounded on the impacts of tourism, which has raised awareness of the negative environmental aspects that can result from its development, while the positive environmental aspects have been less well-defined.

The negative impacts

Resource and ecosystem pressures

Major concerns associated with the impact of tourism relate to the overuse of natural resources and pressures placed on ecosystems. As to how tourism can result in an overuse of natural resources it is useful to consider Hardin's (1968) 'Tragedy of the Commons' parable. This is one of the most cited writings in environmental studies, questioning the assumption of classic economics that behaviour driven by self-interest automatically acts for the greater social good. Using the analogy of an area of common land termed 'the commons' on which farmers in a village are at liberty to freely graze their cattle, Hardin suggests that an existing state of equilibrium between the numbers of cows grazing on it and its ability to regenerate itself can be threatened by the self-interest of the farmers. Specifically, one farmer may decide he wants to increase his herd's milk production and profits by the addition of an extra cow to his herd. While this one extra cow may not directly threaten the commons' long-term stability, the other farmers witness this action and decide that they too would like to increase the size of their herds and their subsequent profits. Ultimately, the pressures placed upon the commons from the extra cattle lead to a situation of over-grazing and the commons becomes unsustainable, as the grass is not permitted to regenerate in sufficient quantities, threatening the long-term existence of the resources and the livelihoods of the farmers that are dependent upon it.

The parallels of Hardin's essay to tourism development are poignant. Entrepreneurs who are attracted by the financial opportunities of serving an emerging tourism market often drive initial tourism development. If the financial return on the investment is proven to be comparatively advantageous, then other investors will also seek to secure the potentially high rates of return through investing in the tourism industry. If environmental planning is weak or absent, then an incremental growth in the number of hotels may occur year upon year. Similar to Hardin's (1968) allegory, the consequential resource and ecosystem pressures originate not from one single hotel, but from the incremental and cumulative effects of the addition of new hotels over a period of years.

The types of common resources that tourism relies upon are typically what in environmental studies are referred to as 'common pool resources' (CPRs). They are characterised by criteria of exclusion and exploitation, where exclusion is impractical on the basis of cost or is at least very costly, and the exploitation of the resource by one person reduces the benefit for another (Ostrom *et al.*, 1999). A major threat to the well-being of CPRs exists from the mentality of 'finders-keepers', that is, a rush to harvest and secure the benefits of the resource before someone else does so (Hardin, 1968). Typical CPRs used for tourism include the oceans and seas, the atmosphere, beaches, coral reefs and mountains. To illustrate how tourism can create pressures that lead to an overuse of resources and threaten their sustainability, a scenario based upon the use of a coral reef for tourism is shown in Figure 7.1. Coral reefs are colonies of animals forming one of the world's most diverse ecosystems, and although they cover only 0.17 per cent of the ocean floor, they are home to approximately 25 per cent of all marine species (Goudie and Viles, 1997). They are also extremely sensitive to environmental changes, requiring highly oxygenated water and a constant temperature of 25–29°C (Goudie and Viles, ibid.).

Figure 7.1 emphasises the reliance of different user groups associated with tourism upon the coral reef to meet their various needs, including financial, economic and hedonistic ones. In this case, the direct user groups of the reef include hotels; local tour operators; tourists; and other enterprises and entrepreneurs. All of these groups would

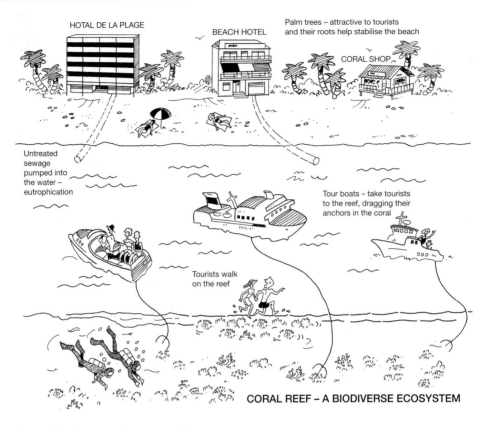

Figure 7.1 The use of a coral reef for tourism.

expect benefits from their interaction with the coral reef. Hotels, shops and local tour operators would seek financial benefits, while tourists would expect to enjoy themselves. At a macro-level, the government would expect economic benefits to be generated by the use of the reef for international tourism. However, the pursuit of benefits may lead to damage to the coral as a consequence of increased demand and inadequate governance and management.

As is suggested in Figure 7.1, a major threat that tourism can pose to the coral is through pollution, especially the release of untreated sewage into the sea. Although untreated sewage is a consequence of human habitation, there is little doubt that tourism compounds the quantity of human waste, especially in the major tourist areas of the world. In the Mediterranean Basin, only 30 per cent of over 700 towns on the coastline treat their sewage before discharging it into the sea, while in the Caribbean Basin only 10 per cent of sewage is treated before discharge (Jenner and Smith, 1992). Compared to other tourist-receiving areas of the world these figures are actually good, with other significant regions such as East Asia, east Africa and the islands of the South Pacific possessing either no sewage treatment or treatment plants that are totally inadequate for the size of the population (Jenner and Smith, ibid.).

A major effect of untreated sewage emission into water is nutrient enrichment, termed 'eutrophication', which stimulates the growth of algae. This can lead to the suffocation of the living organisms that make up the reefs or atolls, as happened off the Hawaiian island of Oahu, with large parts of the reef being killed (Edington and Edington, 1986). Similarly, a proliferation of starfish as a consequence of pollution threatens the ecosystem of the Great Barrier Reef, the world's largest coral reef and a major tourist attraction off the east coast of Australia. According to Jenner and Smith (1992), millions of starfish have been observed, each one able to eat an area of coral the size of the starfish itself in a day, and as much as one third of the Great Barrier Reef has been adversely affected to some degree.

Besides being a natural attraction for tourists, coral reefs form the centre of a wider ecosystem. Within an ecosystem exists a set of complex relationships, which are finally balanced, and similar to the analogy of the tourism system to a spider's web. The ecosystem of which coral is a key part provides many benefits for humans besides its beauty. Coral reefs act as a natural breakwater to help protect coastline and beach areas from erosion, without which erosion would occur at a much quicker rate. The reef ecosystems also include an abundance of fish, for example Pacific islanders get up to 90 per cent of their protein requirements from reef fish (Goudie and Viles, 1997).

Although tourism may pose a significant threat to coral reefs in some localities, for example 73 per cent of the coral reefs off the coast of Egypt are thought to have been adversely affected by tourism (Goudie and Viles, ibid.), it is important to keep tourism's role as a polluter in perspective. At a global level, other factors including global warming, tropical storms, increased sedimentation as a consequence of deforestation, industrial pollution, and over-fishing, pose more significant threats to the reefs than tourism.

The effects of climate change have already been observed on coral and the tourism industry. Global warming affects the patterns of ocean currents and the consequent movement of warm and cold waters around the globe. Owing to the specific environmental conditions for the growth of coral an increase of 1–2 degrees centigrade can result in coral bleaching and dying. Coral bleaching occurs when corals expel the tiny algae or *zooxanthellae* that live with them. These algae are essential for providing nutrients to the living coral polyps. In 1998, the warmest ever year recorded, coral bleaching took place on the resort of Bandos Island in the Republic of Maldives, destroying the coral. The impact on the tourism industry, which is based upon diving, was marked, with a 30 per cent decrease in the normal income recorded from diving (WTO, 2003c). In the same year coral bleaching at Palawan Island in the Philippines led to coral mortality of 30 to 50 per cent. Provided the coral recovers by 2008, the predicted economic effect upon tourism from this damage is estimated to be a loss to the national economy of US$6–7 million. If the recovery takes longer, the predicted loss of earnings will be in the range of US$15–27 million (WTO, ibid.). An example of the more general impacts tourism can have upon coral is described for the Caribbean in Box 7.2.

Think point

Have you observed the negative effects of tourism on the natural environment of any coastal area you have visited? If so, what were they?

Box 7.2 The impacts of tourism upon coral in the Caribbean

The kinds of pressures that coral can be placed under from tourism stakeholders are exemplified by its use for tourism in the Caribbean. These include stakeholders and scuba divers breaking fragile branching coral with their flippers and killing marine life by spear fishing; the dragging of boat and yacht anchors through the coral (US Virgin Isles); dumping garbage from boats onto the reefs (the Grenadines); over-fishing and dynamiting of reefs for lobster and conch to sell to tourists (US Virgin Isles); tourists walking on the coral in plastic sandals (the Buccoo Reef in Tobago); beach vendors selling rare black coral made into earrings (Grenada) and the backs of endangered turtles (Barbados). It is estimated that cruise-ship anchors in the Cayman Islands have destroyed more than 300 acres of coral reefs.

Source: Pattullo (1996)

Besides impacting upon the ecosystems of coastal areas, the second most popular location for holiday tourism is mountain areas. Many mountain areas in the world have become increasingly popular for tourism during the twentieth century, with an increased demand for activity-based tourism, including downhill skiing, mountain biking, hiking or tramping, mountain climbing, paragliding and white-water rafting. However, despite their robust appearance, mountain areas are complex ecosystems that are sensitive to change. They often have short growing seasons as a consequence of cold temperatures and soils that tend to be thin and nutrient deficient.

The development of tourism in mountain areas has brought economic benefits, notably employment opportunities and wealth creation. These are important, as previously many of these areas had been suffering from depopulation and unbalanced demographic structures, as young people left their villages to search for work in urban centres. However, as in the coastal areas, the economic success of tourism creates pressures for further development. The development of tourism in mountain areas requires the construction of a tourism infrastructure of hotels, apartments and associated transport and utility networks, placing increased pressure on land resources and animal habitats. For example, the removal of forests to create downhill ski-runs, an activity that is estimated to account for three to four per cent of total international tourism arrivals (WTO, 1998), besides resulting in a loss of habitat for wildlife, also means that rainfall on the mountain slopes is not absorbed in the same quantity as before. The removal of trees causes a loss of cohesion and stabilisation of the soil by the tree roots, and subsequently the mountain slope is more prone to slippage. The combined effects of increased amounts of water running across its surface, weakened stability and the force of gravity have made mountain areas more prone to landslides. The effects of these landslides can be quite dramatic, sometimes involving the loss of human life. For example, Simons (1988), describes the avalanches of mud that swept down mountainsides during the summer of 1987 in northern Italy and southern Switzerland, in which 60 people died, 7,000 were made homeless and 50 towns and villages and holiday centres were wrecked. The cause of these landslides was attributed to the removal of mountainside forest for ski development.

Figure 7.2 Mountain areas have become popular areas for activity tourism. Despite their robust appearance, mountain areas have finely balanced ecosystems.

The future of winter sports in mountain regions is itself under threat from climatic change. The world's mountain regions and most popular ski areas of the European Alps and the US are being affected in a similar way, as the snowline is receding due to warmer winters (WTO, 2003C). This means that in the future, lower altitude resorts are likely to be less popular and higher ones more popular, placing extra environmental stress on these areas. Consequently, lower altitude resorts may be required to introduce alternative attractions to skiing during the winter season, while environmental management will be an increasing priority for the higher resorts.

Besides ecosystems there are a variety of other types of resources on which tourism can put pressure, including land. The movement of hundreds of millions of people around the world necessitates a complex transport and accommodation structure to support this movement. Airports, seaports, railways and road systems all demand large quantities of land, often transforming it from other economic uses such as agriculture, creating an 'opportunity cost' of lost agricultural production. For example, at Heathrow international airport in England, agricultural land equivalent to 200 miles, or 320 kilometres, of three-line highways has been paved over (Friends of the Earth, 1997). Sometimes, agricultural land loss may lead to a requirement for increased food imports, as in the case of some small islands where land has been lost to airport and seaport developments (Briguglio and Briguglio, 1996).

Besides placing pressures on land use, tourism can also place pressure on another scare resource, water. Some of the world's most water-stressed areas are also recipients of large quantities of international tourism, e.g. the Mediterranean. The addition of

Figure 7.3 The need to economise on water usage may lead hotels to ask guests to restrict their use.

hundreds or thousands of bed spaces in a destination, combined with the lifestyle demands of western tourists, for example a daily requirement for showering, clean sheets and bath towels, means that tourism is responsible in some destinations for the consumption of copious amounts of water compared to the needs of the local population. Besides water required for each hotel room, water is also typically required for general hotel-management activities including the kitchens and laundry, swimming pools, lawns and golf courses (TOI, 2002). Salem (1995) comments that 15,000 cubic metres of water will supply 100 luxury hotel guests for 55 days, 100 nomads or 100 rural farmers for three years, or 100 urban families for two years.

The effects of the development of tourism in areas where water resources are limited can mean that local people are denied the access to the water resources they previously used, for example to irrigate crops. They may find that streams previously used for irrigation have been diverted further upstream to service tourism development, or that the water-table level has been lowered by over-extraction to service tourism establishments, rendering their wells useless. Where it is impossible to continue to extract enough fresh water locally, hotels can pay to have the water imported, while local people continue to suffer water shortages.

Think point

Have you ever been aware as a tourist that water scarcity may be a problem? For example, in your accommodation have you seen information asking you to save water?

Pollution

While natural resources can be subjected to over-extraction, they can also be overused as sinks for pollution. As Hardin (1968: 1245) comments: 'In a reverse way, the tragedy of the commons reappears in problems of pollution. Here it is not a question of taking something out of the commons but of putting something in.' Different categories of pollution are evident in tourism, including water, air, noise and aesthetic pollution. The problems of water pollution from the discharging of untreated sewage were discussed in the preceding section of the chapter in the context of coral reefs. The nutrient enrichment of water may also threaten the viability of tourism destinations besides those in coral-rich areas. For example, the eutrophication of the Adriatic and the consequent spread of algae led to a decrease in tourism demand of 25 per cent in 1989 compared to 1988 on the Romagna coast of Italy (Becheri, 1991). Although in this case discharges from the chemical industry were to blame rather than the tourism industry. The contamination of water by faeces also has consequences for human health, causing diseases ranging from mild stomach upsets to typhoid. Similar to eutrophication, such contamination may threaten the success of the tourism industry. For example, in 1988 the fear of a typhoid outbreak in Salou in Spain resulting from contaminated water led to a 70 per cent decline in tourist bookings the next year (Kirkby, 1996). The failure to treat human waste can result in infection, gastro-intestinal disease, leptospirosis and cholera (TOI, 2002).

Alongside human waste, the use of fertilisers and herbicides on golf courses and hotel gardens may also cause water pollution, running directly off the surface in rainwater into water courses, or seeping through to underground aquifers and eventually reaching rivers, lakes and seas (Mieczkowski, 1995). Other sources of water pollution are caused by motorised leisure activities such as power boating, and even suntan oil being washed off tourists when swimming can result in localised pollution. Hotels also produce significant amounts of 'greywater', from washing machines, sinks, showers and baths, and 'blackwater' from kitchen dishwashing and toilets (TOI, 2002). The lack of treatment of this water in many destinations leads to pollutants such as *fecal coliform* bacteria and other chemicals being discharged directly into the environment.

A second type of common resource that tourism can pollute is the air. The major contribution to this type of pollution is the energy demands of the transport system that is required to move hundreds of millions of tourists. For example, in the US 76.5 per cent of tourism's contribution to greenhouse gas (GHG) emissions comes from transport, with the remainder coming from accommodation, restaurants and retail activities (WTO, 2003C). A reliance on car and air transport causes carbon dioxide (CO_2), the key source of GHG emissions to be released into the atmosphere, contributing to global warming.

Air traffic has grown at an average rate of five to six per cent per annum over the last 50 years, currently accounting for three per cent of the world's CO_2 emissions. The strong demand for air travel has led to air transport being the fastest-growing source of GHG emissions and it is predicted to be contributing in the range of 6–10 per cent of the world's total CO_2 emissions by 2050 (WTO, 2003C). Besides contributing to global warming, air transport also emits two to three per cent of the total global emissions of nitrous oxides (N₂O), which are believed to reduce ozone concentrations in the stratosphere as well as contributing to global warming (Friends of the Earth, 1997). One of the predicted consequences of global warming is a rise in sea level of between 20cm and 100cm during this century. For small islands that has dramatic implications, for example in the Maldives a large proportion of the land mass could disappear, and saltwater intrusions into freshwater aquifers could make the islands uninhabitable (WTO, ibid.). Tourism also contributes to local and global air pollution through the energy requirements of hotels and resorts, the majority of which meet their energy needs through purchasing energy produced from the burning of carbon fuels, i.e. coal, oil and natural gas (TOI, 2002).

The development of airports also has been found to impact upon health, including respiratory problems caused by emissions from aircraft and car traffic, and stress associated with noise pollution from air traffic. According to Whitelegg (1999), aircraft produce significant amounts of nitrous oxides during takeoff and landings, adding that Kennedy and La Guardia airports in New York are among the largest source of polluters in the city, and that Midway Airport in Chicago generates more toxic pollutants than any other form of industry in the city. The effects of this pollution on health are marked, with aircraft engines being held responsible for 10.5 per cent of the cancer cases in southwest Chicago caused by toxic air pollution (Whitelegg, ibid.).

While air travel causes significant environmental impacts, the most common means of transport for tourism is the car. The pollution, congestion and road safety problems associated with car transport are among the major ones facing many urban societies in the world. These problems are replicated in many tourism destinations particularly during peak seasons of the year.

Figure 7.4 Pollution, congestion and road safety problems associated with car use can be major problems in some destinations.

People living in tourism transit areas may also be subjected to pollution problems that could impact upon their health. For example, commenting upon the impact of transit traffic through the European Alps, Zimmermann (1995: 36) writes: 'The transit traffic is one of the most evident problems within the Alpine area. In several regions local populations' endurance levels have already been reached or exceeded.' Associated with transport issues, noise pollution may also be significant from increased levels of traffic, especially for people living near principal car and air routes. This includes people living near airports and under the flightpaths of aircraft. Noise pollution may also be caused by the building of hotels and other types of construction activity in holiday destinations (Briguglio and Briguglio, 1996). Nightclubs open until the early morning and increased car traffic from tourism movements all add to the noise pollution experienced by both residents and tourists in tourism destinations.

Perhaps one of the most evident aspects of pollution associated with tourism is the aesthetic transformation of tourist areas. One conspicuous feature is the urbanisation of coastlines in heavily developed tourism regions of the world. The typical concerns of such urbanisation are summated by Burac (1996: 71) in reference to Guadeloupe and Martinique: 'The most worrying problem now prevalent in the islands relates to the anarchic urbanisation of the coasts. . . . Also, the built-up areas by the seaside are often not aesthetically attractive due to the diversity of architectural styles.' Unfortunately from an aesthetic viewpoint, tourism development has been frequently based upon the maximisation of profits rather than paying attention to aesthetic concerns. This has led to a uniform style of development along many coastlines of the world popular for tourism, ignoring local architectural styles, building traditions and materials resulting in a visual similarity.

In mountain areas tourism has also created unsightly development. Besides hotel and apartment construction, the development of ski lifts and pistes has been heavily criticised as a form of aesthetic pollution. For instance, the Scottish Office (1996: 7) make the following remarks about the development of downhill ski facilities in Scotland: 'In addition [to adverse ecological effect], the infrastructure and uplift facilities associated with skiing can have a visual impact on what would otherwise be an unspoilt and undeveloped landscape.'

Think point

Tourism can have negative consequences for resources and cause pollution. Yet, tourism can also bring economic prosperity and social stability to areas that were suffering from economic and social decline or in poverty. To what extent are the loss of natural resources and pollution important if economic benefits are being brought to an area from tourism?

Environmental quality and tourism

Pollution and resource depletion, besides causing an ecological disequilibrium, also threaten the attractiveness of destinations. For most destinations a high level of environmental quality is essential to make them attractive to tourists, as Mieczkowski (1995:

114) comments: 'The very existence of tourism is unthinkable without a healthy and pleasant environment, with well preserved landscapes and harmony between people and nature.' This point is further supported by a survey of German tourists that found that the requirement for a beautiful landscape was the single most important factor for a 'quality tourism' destination (European Tourism Analysis, 1993).

The environmental threats that tourism can pose represent one of the dilemmas and paradoxes of tourism, summed up in the well-rehearsed metaphor 'killing the goose that laid the golden egg'. That is, although it is often the natural beauty of an area that attracts tourists, development pressures and increased numbers of visitors can destroy the beauty of the nature that attracts tourists.

The idea of how a destination may transform and develop over time, and its natural resources be placed under pressure and transformed, can be explained through what is commonly referred to as the 'destination life cycle' proposed by Butler (1980). The concept of the 'hypothetical evolution of a tourist area' as Butler (1980: 7) termed it is shown in Figure 7.5.

The implications of this model are that for a hypothetical destination, different stages of development can be observed. The first is the 'exploration' stage in which tourists are attracted by the 'otherness' of the destination's natural and cultural features with an emphasis being on their 'unspoilt' characteristics. These kinds of tourists may be likened

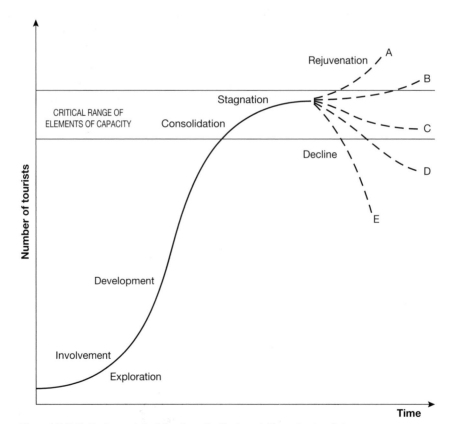

Figure 7.5 Butler's model of the hypothetical evolution of a tourist area.

to Plog's 'allocentrics' described in Chapter 2. At this stage there is no special provision of facilities for tourists. The 'involvement' stage is when a more formalised contact between local people and tourists can be recognised, as facilities including catering and food establishments are provided exclusively or primarily for tourists. A pattern of visitation based upon some kind of tourist season may also emerge. The 'development' stage marks the existence of a well-defined tourism market, with heavy advertising of the destination taking place in tourism-generating areas. At this stage, changes in the physical attributes of a destination will be evident and the natural and cultural attractions will be developed and marketed to the tourists. The type of tourist will have changed, typically arriving with tour operators, typifying Plog's 'midcentric'. At the 'consolidation' stage the rate of increase in the numbers of visitors will begin to decline, with a major part of the area's economy being dependent upon tourism, and the land-use of the destination being heavily weighted to tourism. In the 'stagnation' stage, Butler (1980: 8) comments: 'The resort image becomes divorced from its geographic environment', as its natural attractions have been replaced by attractions of a more human-made and artificial type. The type of tourist coming to the destination is likely to be of Plog's 'psychocentric' types. From this stage the resort either will go into a stage of 'stagnation', unable to compete with new destinations in the marketplace, or will find some means for 'rejuvenation', for example by finding new types of markets, e.g. meetings, incentives, conventions and exhibitions (MICE), or gambling.

As Butler (1980) points out, this sequence of development is not a pre-determined cycle of evolution for a tourist area. However, in terms of a change in the natural characteristics of the destination, the implication of the model in each stage is that there is a loss of quality in the natural environment, as the destination becomes more developed and more reliant upon human-made attractions. While the types of tourists coming to the destination changes with time, the examples of Salou in Spain and the Italian Adriatic, given earlier in the chapter, suggest that without careful environmental planning and management, at some point a 'shear point' is inevitable. The shear point is where the environmental quality of a destination is viewed by the market to have reached such a poor quality that it begins to lose its market share and decline. As Jenner and Smith (1992: 178) comment:

> There comes a point, often suddenly and seemingly unpredictable, where tourism comes to a virtual full stop due to some severe environmental problem. It could be algal blooms, as in the Adriatic, which have almost literally prevented entry into the water. It could be an oil spill on the beaches.

Think point

Looking at travel supplement in a paper, how are nature and wildlife promoted to tourists? What do the advertisements suggest about how the environmental ethic of society has changed since the early twentieth century in terms of our interaction with wildlife?

Positive impacts

It is debatable whether any kind of human activity based upon the use of natural resources can have a beneficial effect for the environment, other than to protect it from more damaging forms of human behaviour. Therefore when we talk about the positive environmental impacts of tourism, we are in essence talking about tourism being used as a way of protecting the natural environment from potentially more damaging forms of development activity, like logging or mining. While landscapes can have a range of values attributed to them, including a life-support value, recreational value, scientific value, aesthetic value, genetic-diversity value, historical value and religious value (Holmes Rolston III, 1988), it is often their economic value that is decisive in deciding their use.

It is the willingness of tourists to pay for travel and tourism services to experience nature and observe wildlife, transforming its aesthetic and recreation value into an economic one, which provides the rationale for conservation. A significant niche market of this genre is wildlife tourism, and the changing focus and purpose of wildlife tourism illustrates how environmental attitudes in society can influence the relationship between tourism and conservation.

While wildlife tourism is commonly associated with the 'viewing' of wildlife this was not always the case. In one of the most popular areas of the world for wildlife tourism, the east of Africa, at the beginning of the twentieth century safari-tourism was based upon the 'shooting' rather than the viewing of animals. This activity was participated in only by the very wealthy on vast colonial estates. The scale of devastation upon the numbers and diversity of species of animals as a consequence of game shooting was enormous. For example, within a few decades of the Europeans arriving in the east of Africa the blaubok and quagga were eliminated, both of whom had survived three million years of contact with the indigenous people. One notable example of the carnage caused through hunting was an expedition led by Theodore Roosevelt and his son, in which 5,000 animals of 70 different species were killed, including nine of east Africa's remaining white rhinos (Monbiot, 1995). For many readers the scale of this carnage may seem shocking, and it is this change in the environmental ethic of many societies towards wildlife that helps to explain why emphasis is now mostly placed upon observing animals rather than killing them.

Think point

Looking in a travel supplement at a paper, how is nature and wildlife promoted to tourists? What do the advertisements suggest about how the environmental ethic of society has changed since the early twentieth century in terms of our interaction with wildlife?

It is the willingness of international tourists to pay to view wildlife and natural habitats that provides a strong economic incentive for governments to act to conserve natural resources. Today, the diversity of wildlife-watching ranges from the viewing of raptors in the north of Greece, gorillas in Rwanda, the safari game of the east of Africa, and whale watching off the Patagonian coast of Argentina, to give a few examples. One method of conserving nature and wildlife is through the establishment of protected areas, most notably, with relevance to tourism, through the creation of national parks. The establishment of national parks in less-developed countries acts as an important focal point for attracting international tourists, for example as in east and south Africa, Costa Rica, India, Nepal and Indonesia (World Tourism Organisation, 1992).

Although national parks have an important function in less-developed countries for conserving wildlife and nature, the first national park was created in Yosemite in the US in 1890. Highly influential in its creation was John Muir, who championed the conservation of nature. Muir viewed nature as being valuable for people, in terms of recuperation, aesthetic appreciation and spiritual replenishment (Nash, 1989). It is not a coincidence that the founding of Yosemite coincided with the increasing industrialisation and urbanisation of America, in the latter half of the nineteenth century. Hall and Lew (1998) suggest that the development of the Yosemite National Park represented a need for a rapidly urbanising American population to stay in contact with nature. More polemically, in the view of MacCannell (1992: 115), national parks: 'are symptomatic of guilt which accompanies the impulse to destroy nature'.

Criticisms of the establishment of national parks in less-developed countries have also been made on the basis of their exclusivity of local people as Leech (2002: 78) comments:

> Though many of us crave sanctuaries away from the modern world, the idea of wilderness areas where tourists can sample nature, free of man, is a western romantic illusion. And further, what is evident is that the preservation of these areas comes at a very high cost to local communities which are either removed from the land, have their lives regulated or are forced to play the role of theme park extras to satisfy the demands of ecotourism.

Many national parks are created in already populated land, for example in Latin America it is estimated that 86 per cent of protected areas were already populated (Leech, ibid.) One of the most notable examples of the displacement of indigenous people associated with tourism was the exclusion of the Maasi people from their traditional lands, when the Maasai Mara reserve was established in Kenya in the 1940s. This has radically altered their way of life and resulted in some of the Maasi having to migrate to the coast to sell their handicrafts to tourists. However, the exclusion and displacement of local people from lands for tourism development is not something that is specific to Kenya. The development of the Chitwan National Game Reserve in the Terai area of Nepal was also achieved by the exclusion of local people.

Think point

Should national parks be created on a 'preservation and exclusion ethic' if this means the displacing of people who inhabit the area?

Nevertheless, the economic rationale for nature conservation will be even stronger when tourism is calculated to have a potentially greater economic value than other development options. In less-developed countries, often pressured to earn foreign currencies to service foreign debts and to boost exports, nature tourism may offer an attractive method of earning foreign exchange. The earning power of wildlife tourism compared to other development options is demonstrated through research in the Amboseli National Park, in which wildlife tourism was found to be economically preferential to the other main development option of agriculture. The park's net earnings from tourism were found to be US$40 per hectare per year, 50 times higher than for the most optimistic projection for agricultural use (Boo, 1990). However, although the argument for conservation based upon nature's economic value is persuasive, it should be treated with caution, as the counter side to this argument is that if a species does not possess a high enough economic value it should not be conserved.

Think point

Should nature and wildlife be conserved only if it has an economic value? Are there other reasons to conserve nature? If so, what are they?

An ethical use of the environment for tourism?

A key determinant of how tourism interacts with the environment is the environmental ethic held by its stakeholders, including tourists, tourism businesses, local communities and governments. The value humans give to nature and how we interact with our natural surroundings is of key interest in environmental studies. Significant in this line of thinking is the work of the Norwegian environmental philosopher Aerne Naess (1973). He typified two broad approaches to our interaction with nature based upon the value we give to it, termed 'shallow ecology' and 'deep ecology'. Shallow ecology is based upon an 'anthropocentric' view of nature, meaning that nature is viewed as being separate from humans, and its value rests purely in terms of the use it has in meeting human needs and desires. Consequently, the anthropocentric view of why the environment should be conserved or treated in a responsible way rests solely with the benefits this would bring for humans.

In contrast, 'deep ecology' rejects any separation of nature and humanity, stressing their inter-connectivity, and that all beings are of equal value. A value is given to nature, which emphasises its right to existence, rather than its instrumental value. Thus, rather than assuming that society should utilise natural resources for its own benefit, deep ecologists would question the purpose of the use of those resources and whether they were really necessary or not. Deep ecology subsequently challenges the values of a capitalist- and consumer-based society, emphasising that: 'society–nature relationships cannot be fundamentally transformed within the existing social structures' (Pepper, 1996: 21). Subsequently, deep ecology stresses the requirement for social change based upon the transformation of 'individual consciousness' (Pepper, ibid.) Thus there is a need for individuals to adopt an environmental ethic that is reflected in their lifestyle and behaviour that emphasises a respect for nature.

Although Naess's ideas are significant, the questioning of our interaction with nature can be traced to more than a century earlier and the works of Henry Thoreau (1817–1863). Living at a time when America was rapidly industrialising, Thoreau observed associated changes in his home town of Concord including the arrival of the railway, the disappearance of familiar animal species such as the beaver and deer because of over-hunting, and the removal for use as fuel of much of the local forest he grew up with (Walls, 2001). He was highly critical of the way that capitalism encroached upon previously open land, enclosed it and fenced it off against trespassers. Written during his two-year stay in comparative isolation at Walden Pond, a glacial lake close to Concord, his most famous work *Walden*, published in 1854, questions how nature is used for capitalism. The work was influential in recognising a deeper valuation of nature that went beyond its instrumental value and was influential in the later founding of the American environmental movement.

In a similar vein to Thoreau, though writing nearly a century later, Aldo Leopold (1886–1948) also stressed the need for a more integrated and holistic relationship between humans and nature. Advocating the need for a 'land ethic', in his most famous work *A Sand County Almanac*, Leopold (1949: 219) comments: 'In short, a land ethic changes the role of Homo Sapiens from conqueror of the land-community to plain member and citizen of it. It implies respect for his fellow-members, and also respect for the community as such.' Leopold was suspicious of recreation and tourism, viewing it as already having a negative impact on wildlife, and the travel trade as encouraging this trend. Besides their impact upon nature, he was particularly concerned about how the tools of advertising and promotion in the 1940s were being used to inspire access to nature in bulk, and consequently reduce the opportunities for solitude (Hollinshead, 1990). This idea of inseparability between nature and humanity was taken to its most holistic perspective by Lovelock (1979) in his concept of 'Gaia', which argues that the earth should be viewed as a single living organism composed of different and interrelating and regulating parts, one of which is humanity.

Although the ideas of Ness, Thurot, Leopold and Lovelock may seem rather esoteric or abstract, they have a direct relevance to the future of tourism. Not least, these different perspectives give rise to a variety of interesting reflections and questions. These include: should the environment be used in an 'instrumental' way to maximise economic benefits and financial profits of tourism; can environmental management and technologies solve the environmental problems that can be created through tourism; do we need to have a stronger environmental ethic that recognises an independent value of nature?

There is evidence that an ethical emphasis is increasingly influencing the action of tourism stakeholders on the basis that some existing patterns of behaviour can harm the environment. For example, codes of conduct have been developed by various organisations in the private and public sectors to help guide appropriate tourist behaviour. The private sector seems to be placing an increased emphasis upon demonstrating corporate social responsibility (CSR), and what is commonly referred to as the 'triple bottom line' to assess a company's performance, incorporating the social, environment and economic impacts of a business's operations besides purely the financial, as was highlighted in Boxes 5.4 and 5.5.

The behaviour of tourists as consumers is also being challenged with calls for less holiday taking and the taking of holidays closer to home. Although such calls may have previously seemed marginal to the mainstream thinking of society, often originating from NGOs, they are now finding their way into more popular outlets. For example, in an article in a popular British newspaper titled '30 Ways to do the Right Thing', which was aimed at creating a more environmentally friendly society, one recommended way was:

> Ration Your Flights. Air travel is the world's fastest growing source of greenhouse gas emissions–one return flight to Miami creates more carbon dioxide production than one year's motoring. One million people in the UK are adversely affected by aviation noise and air pollution, and expansion uses up vast swathes of countryside. Ration yourself to one flight a year, take the train and ferry, or holiday in the UK.
>
> (Siegle and Templeton, 2004: 37)

Think point

Would you be willing to modify your pattern of behaviour as a tourist to protect the environment, for example by taking less flights?

Summary

- Recognition of the potential negative effects of human activity upon the environment grew from the 1960s. The first widely broadcast images of earth as a sphere floating in space, sent from *Apollo 8*, created awareness that 'spaceship earth' did not have an unlimited abundance of natural resources. Today, environmental issues such as pollution, global warming and ozone depletion have become a part of the global lexicon.
- The beginnings of an awareness that tourism could have negative effects upon nature coincided with the development of mass international tourism during the 1960s and 1970s. Observations of the destruction of nature and culture by tourism began to be voiced. As the spatial boundaries of mass tourism expanded during the 1980s to include less-developed countries, NGOs campaigning for a type of tourism more respectful of nature and other cultures emerged, notably the Ecotourism Society in the US and Tourism Concern in the UK. At the end of the 1990s tourism-militant environmentalists attacked facilities for the first time at Vail in Colorado. Also, during the 1990s, the tourism industry began to take action over the environment.
- Tourism can induce change in nature but can also be placed at risk by changes in nature, notably global warming and pollution. Parallels between the 'Tragedy of the Commons' and the potential risk tourism can place on common natural

resources is poignant. Typical types of common resources used for tourism include the oceans and the seas, the atmosphere, beaches, coral reefs and mountains.

- Tourism can give an economic value to nature that provides a strong argument for its conservation. Although nature has a range of values, including the recreational, aesthetic, scientific and historical, development decisions are usually based upon its economic value. It is the willingness of tourists to pay for travel and tourism services to experience nature and wildlife, giving it an economic value, which provides a stronger rationale for conservation. This economic rationale may provide a motive for governments to establish protected areas for the conservation of nature and wildlife, particularly within the context of tourism: national parks. However, although the argument for conservation based upon nature's economic value is persuasive, it should be treated with caution, as the counter side to this argument is that if a species does not possess a high enough economic value it should not be conserved.

- Influential in determining how tourism interacts with nature will be the environmental ethic of its stakeholders. Does nature have purely an instrumental value or does it have an independent value and right to existence?

Suggested reading

Holden, A. (2000) *Environment and Tourism*, Routledge, London.
Hunter, C. and Green, H. (1995) *Tourism and the Environment: A Sustainable Relationship?*, Routledge, London.
Mathieson, A. and Wall, G. (1982) *Tourism: Economic, Physical and Social Impacts*, Longman, Harlow.
Mieczkowski, Z. (1995) *Environmental Issues of Tourism and Recreation*, University Press of America, Lanham, MD.

Suggested websites

United Nations Environment Programme   www.unep.org
United Nations Educational, Scientific and Cultural Organisation   www.unesco.org

EMERGING THEMES OF TOURISM

8

This chapter will:

- introduce ethical theory;
- describe ethical issues of the tourism system;
- consider how ethical theory can be used to evaluate these issues;
- introduce theoretical perspectives of feminist studies;
- consider issues of women's empowerment and disempowerment;
- explore feminist interpretations of these issues.

Ethics and tourism

It may seem odd to think of tourism as an ethical issue like, for instance, abortion, but the complexity of tourism brings with it a range of ethical questions. From the discussion in previous chapters it is evident that tourism raises issues of the equality and type of interaction both between humans and with our surroundings. Equality, fairness, morality, the rights of animals and nature are all relevant issues to tourism. Given that tourism is predominantly based upon meeting the needs of individuals who are usually paying for the privilege, Smith and Duffy (2003: 7) ask some basic questions about the ethics of tourism:

> Is tourism all about the egoistic satisfaction of those paying for the privilege or should ethics play a part? What does it mean to say that a certain way of behaving, or a particular kind of tourism development, is wrong? Can the tourism industry 'afford' morality?

Other examples of ethical questions that could be asked about tourism include: is participation in sex tourism wrong?; is the shooting or killing of wildlife justified?; do tourism multinationals have ethical responsibilities to the communities and nature in which they operate?

For these questions to be even considered represents a cultural change in how tourism is viewed by society. Commenting on the recent emergence of ethical thinking to tourism, Fennell (2000: 59) states: 'Only in the past decade have tourism researchers begun to consider the value of linking tourism and ethics.' In his view it is not a question of 'whether' ethics will be incorporated into tourism studies and research but merely a question of 'when'. For both Fennell (ibid.) and implicitly for Lea (1993), the pressing ethical dilemmas of tourism include exploitation of the developing world, women and nature. It is subsequently possible to view the application of ethics to tourism as a continuation of impact studies, concerned with 'what ought to be' in place of 'what is'.

Writing at the end of the 1990s on behalf of Tourism Concern, an NGO campaigning for more ethical tourism referred to in the last chapter, Wheat (1999a: 3) emphasises the rationale for ethics in tourism: 'It has taken a lot of work over the years to get people to understand that "sustainable tourism" included people as well as wildlife. Which is how 10 years on [from Tourism Concern's foundation] we have moved to the term "ethics".' Alluding to the ethical issues concerned with the impacts of tourism, she (ibid.) continues:

> The tourism industry has grown rapidly, but poverty has increased and people have started questioning why, when so much money is allocated to tourism, so little of it is getting to the people who need it most. Countries with harsh political regimes have increasingly turned to tourism to line their pockets, buy arms and legitimise their governments. Tourism resorts and all the facilities that they require, have time and again taken land and natural resources away from local communities.

While for many stakeholders in tourism these views may be contestable, it would be naive to suggest that a system as complex as tourism did not require a debate about the ethics and behaviour of its stakeholders. The need for debate and the diversity of ethical issues of tourism has been recognised by the World Tourism Organisation who at the end of the last decade published their *Global Code of Ethics for Tourism* (WTO, 1999). The code is an extensive one of 16 pages and includes issues of 'mutual understanding and respect between peoples and societies'; 'sustainable development'; 'cultural enhancement'; 'benefits for host countries and communities'; 'the right to tourism'; 'liberty of tourist movements'; 'rights of workers and entrepreneurs in the tourism industry'; and the 'duties of stakeholders'.

Ethical theory

To support an opinion about the ethics of tourism it is necessary to have knowledge of mainstream ethical theory and morality. Broadly, ethics may be thought of as the branch of philosophy that deals with moral problems and judgements (Fennell, 2000). It is therefore concerned with what is 'right' or 'wrong' and the morality of an action. Consequently, a key question facing philosophers is where does morality come from? This is a problem that has been considered at least as far back as the civilisation of Ancient Greece and almost certainly before then.

An obvious response may be to suggest that morality comes from religion, as a form of puritanical prohibition (Singer, 1993). Most religions provide a system of religious rules to be obeyed, for example Christianity gives the Ten Commandments to obey. In reality we know that these are difficult to always follow and often are open to interpretation in different contexts. For example, the majority of people in most societies would agree with the command 'Thou shall not kill', yet in certain contexts, such as national wars, religious leaders and the state might condone aggression and killing. Similarly, if our family was threatened and the only choice was to kill the assailant or have our family killed, the majority of us would probably feel morally justified to kill the assailant. Additionally, based upon the increased rates of vegetarianism in many western societies and calls for animal rights, it can also be asked if the killing of animals is morally justified.

Singer (1993) prefers to treat ethics as being completely independent of religion, despite the fact that some theologians stress that ethics cannot do without religion, as the meaning of 'good' is what 'God approves'. He comments:

> Plato refuted a similar claim more than two thousand years ago by arguing that if the gods approve of some actions it must be because these actions are good, in which case it cannot be gods' approval that makes them good. For example, if God condoned torture and rape, would this mean that torture and rape were good?
>
> (1993: 3)

Also, many people who do not believe in God, i.e. atheists and agnostics, seem to act morally and ethically. Such issues suggest that instead of there being moral absolutes, morals may be subject to change and varying interpretations in different contexts, and subsequently what is regarded as being ethical behaviour may also change by situation. The view that the beliefs and values of different cultures are varied but are ethically valid is termed 'moral relativism'. Although moral relativism may seem a more logical option than there being some kind of 'moral absolutism', against which to evaluate how ethical our actions are, it potentially poses some testing dilemmas. For example, if a culture condones public hanging and torture does this make it ethically correct? Taking a position of moral relativism may also pose some interesting dilemmas for the tourist. Does the vegetarian who finds him or herself in a meat-eating culture and is offered meat by their host, compromise their beliefs out of necessity or a wish to not cause cultural offence? Does the feminist visiting a country that has a strong patriarchal culture in which women cover their heads and legs, accept this way of dressing even though it may be strongly opposed to her belief system? (Smith and Duffy, 2003).

For the ancient Greek philosopher Socrates (470–399 BC), morality is not the sort of knowledge that you can actually be taught; it ultimately has to be discovered by yourself (Robinson and Garratt, 1999). His concern rested with the soul, believing that to do good is to benefit one's own soul and to do wrong is to harm it (Gottileb, 2000). The outcome of this path of self-discovery is that once we know what is right, we will never do wrong. This implies a high degree of emotional and intellectual awareness to logically guide our behaviour. Yet, we know from our own actions and by talking to others that humans sometimes still do wrong, even when aware that their actions are wrong.

For Socrates' most famous student, Plato, often regarded as the founding father of western philosophy, 'moral facts' existed in coded forms in the universe that could only be interpreted by 'experts' (Robinson and Garratt, 1999). Although this may seem a strange idea, Plato proposed a 'Theory of Forms', which suggested that all material objects and concepts such as beauty, morality and justice were conditions of unchanging forms or blueprints that existed in the universe. Consequently, Plato believed that if experts could comprehend the 'moral facts' that existed in a coded form in the universe, we would know what was 'right' or 'wrong'. Plato was therefore a 'moral absolutist', believing that objective moral truths and facts existed (Robinson and Garratt, ibid.).

The last of the trilogy of renowned Ancient Greek philosophers, Aristotle (384–322 BC), differed fundamentally in his view of morality. He emphasised the requirement of ordinary people to determine their ethics through 'common sense', rather than having them determined for them by an ascetic elite who could understand the 'Forms'. The emphasis in Aristotle's reasoning lies in the conception of purpose, meaning all things could be explained by the purpose they serve, a perspective known as the 'teleological view'. Aristotle emphasised that behaviour is directed towards some final purpose ('telos') or goal (Stokes, 2002). Subsequently, Aristotle's definition of 'good' was determined by the extent to which something fulfils its purpose, e.g. a good knife is one that cuts well, a good holiday is one that people enjoy and return from feeling healthy. This rationale formed the basis of what is called the 'natural law' approach to ethics, in which everything has a natural purpose in life, and actions are right or wrong depending on the extent to which they fulfil that purpose (Thompson, 1999). A summary of this famous trilogy of Greek philosophers is shown in Figure 8.1.

The differing viewpoints of the ancient Greek philosophers emphasise the difficulty of trying to determine the origins and meaning of morality and ethical behaviour. Subsequent philosophical thought has led to the development of a variety of ethical

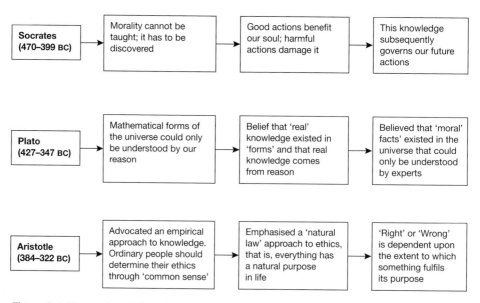

Figure 8.1 The ancient philosophers' views of morality.

theories. One theory developed by Thomas Hobbes (1588–1679) views human nature as being driven by self-interest or 'psychological egoism'. He believed that individuals would always act in their own self-interests, regardless of the interests of others, unless through the consideration of other peoples' interests there are benefits for the individual (Walker, 2000). This means that ultimately we have to compromise because if everyone pursued their own self-interest there would be continuous conflict and social strife, which would threaten their own well-being. Thus Hobbes believed that morality was little more than a method for self-interested but rational human beings to avoid conflict (Robinson and Garratt, 1999).

However, as Robinson and Garratt (1999) point out, Hobbes' explanation of morality originating solely from self-interest is not totally convincing, as many people display behaviour that cannot be associated with psychological egoism, for example jumping into frozen lakes to save children. In an attempt to judge the morality of individual actions, Jeremy Bentham (1748–1832) proposed the theory of 'Utilitarianism', suggesting that an action is likely to be 'right' if it benefits many more people than just the actor (Walker, 2000). Thus, in a situation where there is a moral choice to be made, the correct option is the one that is likely to produce the 'greatest happiness for the greatest number of people' (Thompson, 1999: 71).

For Bentham, the adage of the 'greatest happiness of the greatest number' was the foundation of morals. He also believed that the concept of 'happiness' could be measured in terms of: 'its duration; its intensity; how near, immediate and certain it is, how free it is from pain; and whether or not it is likely to lead on to further pleasure' (Thompson, 1999: 75). Unlike egoism, the emphasis of utilitarianism rests upon the judgement of actions by the 'results' they achieve, not the motives for doing them. The rationale for this is that it is easier to measure or assess the consequences of peoples' actions than their motives. An example of how the concept could work practically is shown in Box 8.1.

However, the use of utilitarianism to decide policy on tourism has its limitations. First, it assumes that we can predict the consequences of actions, for example in the case in Box 8.1 that local people will actually benefit economically from the shooting of wildlife by tourists. Second, the inherent democracy of utilitarianism depends on who is being asked. For example, if sex tourism is proved to produce a greater overall happiness for male tourists than it does for the local population or prostitutes, does this make it morally correct?

The principle of 'Utilitarianism' was refined by John Stuart Mill (1806–1873) who, while like Bentham, advocated that the basic premise of moral action should be the maximisation of happiness or the 'Greatest Happiness Principle', distinguished between higher and lower pleasures (Stokes, 2002). Bentham associated higher pleasures with those of the mind and spirituality, while lower pleasures were associated with the sensual, with a greater value being placed on the higher pleasures rather than lower ones.

Mill further devised two main categories of utilitarianism, called 'Act Utilitarianism' and 'Rule Utilitarianism'. The former of these categories requires that we consider fully the possible outcome of an action to judge whether it is right or wrong while the latter places emphasis on adherence to rules. Act utilitarianism is centred on the form

Box 8.1 Principle of Utility

Bentham constructed a 'felicific calculus', intended to work out the exact quantity of pain and pleasure that would result from a given action. For example, a government wishes to permit wildlife shooting as a form of tourism in a particular area, with the promise of economic benefits for local people. The local people in the area are asked to express their feelings and opinions on the subject based upon the scale below.

Pleasure and pain units:
+ X   = This will make me mildly happy.
+ 2X   = This will make me quite happy.
+ 3X   = This will make me very happy.
+ 4X   = This will make me ecstatically happy.
– X   = This will make me mildly unhappy.
– 2X   = This will make me moderately unhappy.
– 3X   = This will make me very unhappy indeed.
– 4X   = This will make me suicidal.

If the opinion poll results are –3.5 million X units of unhappiness but +5 million X units of happiness, then the wildlife shooting should be allowed. The majority get what they want because Utilitarianism is democratic.

After: Robinson and Garratt (1999)

of argument already outlined, i.e. there are no general rules except that one should seek the greatest happiness of the greatest number (Thompson, 1999). However, Mill doubted the ability of 'ordinary' people to evaluate the possible outcomes of their actions, therefore emphasising that they adhere to rules, a principle he called 'Rule Utilitarianism'.

Not all philosophers accept the principle of utilitarianism, including Immanuel Kant (1724–1804) who is often regarded as the creator of modern philosophy. In his most important written work the *Critique of Pure Reason* published in 1788, Kant claimed to have discovered a universal moral law, which he termed 'the categorical imperative' (Stokes, 2002). The basis of the categorical imperative is that it tells us what we 'ought' to do, regardless of the likely outcome. To help understand what distinguishes a moral action from a non-moral action, Kant developed the 'Universability Test', i.e. 'what would be the consequences of our action if it was universalised?' Another way of thinking of this is the often-heard moral remonstration, 'what if everyone did that?'.

The logic of this principle is that if we are content for everyone else to also do what we are considering doing, it is right, if we are not then it is contradictory and wrong (Thompson, 1999). To help illustrate this point, say we wanted to steal something because we liked it. Although we may be happy to have the object, if by considering our action in terms of 'what if everyone stole from everyone else?' we reached the conclusion that society would descend into anarchy, our action would be judged wrong. The above example also raises the issue of logic about our actions. If we think of our action as being wrong, and think of it becoming a universal law pursued by everyone, it becomes nonsensical. For example, if everyone stole, the concept of stealing would become illogical, as would the meaning of a promise if everyone went around breaking them (Stokes, 2002). From this brief discussion of some of the theories of morality, it is evident that attempting to understand the morality of what is 'right' or 'wrong', and the rules to base our actions upon, is a complex and different undertaking.

Ethical perspectives applied to tourism

That tourism can give rise to a variety of ethical issues is not that surprising considering the variety of stakeholders involved and the possible interactions between them, as is suggested in Figure 8.2.

Most of these interactions are two-way or reciprocal relationships and encompass power dimensions that may leave one party more vulnerable to exploitation or abuse. For example, as was discussed in Chapter 6, one cause of ethical concern relates to the interaction between tourists and local communities. Typical issues include a lack of respect for local cultures and traditions, inappropriate social behaviour including participation in child and sex tourism, and acculturation. Conversely, antagonism and aggression by members of local communities to tourists also raises ethical issues, as do the practices of businesses that attempt to exploit tourists by overcharging or providing poor-quality services. Issues of conditions of employment, including workers' rights, equal pay between men and women, and the issue of trade union representation also fall within the remit of business ethics.

How governments use tourism for economic development may also be ethically debatable. For instance, the use of resources, including cultures, may be at the expense of democratic freedoms, as was exemplified in the case of Malaysia in Box 6.2. Similarly, Bird (1989) describes how a long-established community on the island of Langkawi had its land compulsorily reclaimed by government for a large-scale tourism development.

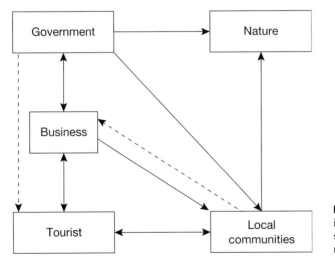

Figure 8.2 The patterns of interaction of tourism stakeholders that may raise ethical concerns.

The extensive development of large-scale golf courses and condominiums in South East Asia has displaced many people and restricted access to essential resources such as water. This pattern has been repeated in other areas of the world such as Goa in India and in North Africa. While the displacement of peoples from their homes may be regarded by some as an unavoidable cost of modernisation and for the greater good, it would seem to be contrary to Article 5(1) of the World Tourism Organisation's Code of Ethics which states: 'Local populations should be associated with tourism activities and share equitably in the economic, social and cultural benefits that they generate, and particularly in the creation of direct and indirect jobs resulting for them' (WTO, 1999: 10).

Think point

The case study of Malaysia in Box 6.2 and other examples suggest that tourism may involve the loss of individual liberties and rights for a greater overall economic good. Do you think this is 'right' or 'wrong'? What conclusions could be reached from the application of the Principle of Utilitarianism to such cases?

Governments also influence the practices of businesses by passing or not passing legislation for the fair treatment and protection of workers in the tourism industry. The issue of human rights in a wider context has also made tourism the focus of ethical debate, in particular whether tourists should visit and subsequently economically support regimes that refuse to recognise human rights, and that are commonly held to be undemocratic. For example, following the occupation of East Timor in the late 1990s by the Indonesian army, the issue of whether tourists should continue to travel to Indonesia became a contested one. Similarly, the question of whether tourists should visit Myanmar (formerly Burma) owing to the governments denial of democratic rights is also a contested issue as is described in Box 8.2.

While tourism raises a number of ethical issues that are dependent upon human interaction, ethical concerns also reverberate around the interaction between the tourism stakeholders and the natural environment. When considering the ethics of tourism it is impossible to ignore ethical questions about stakeholders' interaction with nature. In the context of tourism Lea (1993: 703) comments: 'The industry will come under increasing scrutiny by environmentalists for obvious reasons', many of which were discussed in Chapter 7.

However, it is not purely the industry that interacts with nature. As was explained in the last chapter, so do tourists, local communities and governments. This interaction subsequently raises questions about a more general human interaction with nature than purely the one it has with industry. Typical questions could include: does nature exist purely for its instrumental use for human benefit?; do animals have rights to an existence independent of those given to them by humans?; is it right to kill animals to protect another species? Tourism raises a number of ethical dilemmas for individuals about how they interact with their surroundings. For example, to reiterate a later point of Chapter 7, in the knowledge that air travel pollutes the atmosphere and contributes to global warming, is it 'right' to carry on flying according to our own wishes, or should we try

Box 8.2 To go or not to go to Indonesia?

This is a question that was posed by Wheat (1999b), in relation to the conflict at that time in East Timor following the invasion of the Indonesian army, which raised questions about the ethics of travelling to Indonesia.

For the first time the travel-writing community took an ethical stance on travelling to a particular destination, with the British Guild of Travel Writers putting out a press release that tourists and tour operators should boycott Indonesia until peace was restored in East Timor. However, Indonesia comprises an extensive archipelago of islands stretching approximately 3,000 miles and comprising the world's largest island group. One of these islands is Bali, whose economy and peoples are highly dependent upon tourism. Therefore boycotting Indonesia as a destination for travel would mean especially boycotting Bali.

Accepting that a tourist boycott would lead to economic hardships for many people on other islands of Indonesia, who have no direct connections or vested interest in the political and military struggles taking place in East Timor, raises questions about the appropriate ethical decision for a tourist.

1 Applying the law of universability, what would be for the 'greatest good' in this case?

Many other countries that have been associated with the denial of human rights included the ex-apartheid regime of South Africa, the military junta of Myanamar (Burma), Pinochet's Chile and Marcos's Philippines. In the case of Myanamar, there was a direct call from Aung San Suu Kyi, the winner of the 1991 Nobel Peace Prize and the democratically elected leader, who was placed under house arrest by the military junta, for tourists to boycott visiting Mynamar. This call was supported by the British government and NGOs, including Tourism Concern and Burma Campaign UK, who called for the publisher of the Lonely Planet guide for Myanamar to withdraw their guide (Jenkins, 2002). They refused to do so, saying that it was for the individual tourist to make up their mind.

2 Is it 'right to travel' to a country known to practise the systematic abuse of human rights?

to take fewer flights? Similarly, ethical questions can also arise over how tourists interact with wildlife, as is described in Box 8.3.

The examples in Box 8.3 raise questions about the ethics of human action towards animals. These include: 'do humans have a right to kill animals?'; 'are the methods of culling important?'; 'do the motives and context of killing animals make a difference?'; 'does the fact that tourists are paying for the experience and gaining pleasure from it influence whether it is ethical or not?'. These types of questions are difficult to evaluate using traditional ethical theories and consequently ethical positions have been developed that specifically relate to the interaction between humans and nature, as shown in Box 8.4.

Box 8.3 'Tourists rush for Kill a Seal Pup Holidays' and 'It's the new sport for tourists: killing baby seals'

These two emotive headlines are based upon the decisions by governments in Canada and Norway to allow tourists to participate in the killing of baby seals as part of annual culling programmes. The justification for this action is that there are too many seals, resulting in a decline in fish stocks. Following the government ruling, tour operators decided to promote holidays based on seal-pup culling. The Canadian decision was taken in the early 1990s, with package holidays being marketed for US$3,000 in America, to buy the 'right' to club seal pups to death on the Newfoundland Ice in the annual culling programme. The packages were popular as Mike Kehoe, the executive director of the Canadian Sealers Association commented: 'People want to come out and kill and it's a good market for us' (Evans, 1993).

In Norway, the involvement of tourists in seal culling was set to begin in January 2005, with one company NorSafari advertising culling trips on the Internet. Jowit and Soldal (2004: 3) comment: 'The company's website shows photos of hunters posing with their kill and offers trips that not only include accommodation and food but help with cutting up and preserving seal carcasses.' Although professional seal hunters have traditionally used clubs, tourists would kill the seals by shooting, a presumably more humane, but expensive way of execution. The Norwegian Fisheries Minister said that the move would restore the ecological balance between fish and seals along Norway's coast, although environmental groups argue that over-fishing is the cause of devastated fish stocks, not the seals.

Box 8.4 Ethical perspectives on the environment

Anthropocentric – Nature possesses no value other than its 'instrumental value' to meet humans' needs and wants. There is no recognition that nature has any intrinsic value or 'right' to existence beyond those that may be granted to it by humans.

Libertarian Extension – Nature is recognised as having 'intrinsic rights' of existence independent of those granted by humans. This would include all animals and animate beings, although some would extend this principle to include individual inanimate objects such as trees and rocks.

Eco-holism – Emphasis is placed on the survival of whole ecosystems rather than individual beings. Consequently, the killing of individual animals is permissible if it is for the benefit of the survival of the species as a whole.

With the continued growth in tourism, its growing economic importance, and interaction with an increasing variety of cultural and physical environments it can be expected that tourism will raise more and more ethical questions in the future. As Fennell (2000) suggests, perhaps people don't want to hear too much about the downside of tourism because they enjoy participating in it too much. Maybe this is correct, and for Butcher

---

Think point

Using various environmental ethical positions, evaluate the 'right' and 'wrongs' of the actions of government, tour operators and tourists in the case studies in Box 8.3.

(2002: 59): 'ethical tourism is a barely concealed slight on the "unethical" package holiday maker'. However, in a world in which there is a trend towards green consumerism and demonstrating corporate social responsibility, stakeholders in the tourism system, including tourists, are increasingly likely to have to justify their actions in the future.

Feminist studies

A second emerging area of tourism studies is the application of concepts and theories from feminist studies to tourism. Accepting that tourism induces varying degrees of social change in societies, an important aspect of this is its effects upon gender relationships. The 'rights' of women is an issue that has consequences not only for the future of society. This includes women not only having equal rights and opportunities to men, but also having a key role in influencing and deciding the future direction of societies, their priorities and values.

Ethical issues of tourism, including sex tourism, the opportunity for women to have meaningful careers in tourism and parity with their male counterparts, and the roles of women as travellers, are all issues that fall within the remit of feminism. All of these themes are associated with a traditional hegemonic power of men in many societies and a common theme of all feminist ideas is a belief that women are disadvantaged in comparison with men (Bryson, 1993). Feminists also share the view that this situation is not natural, i.e. a result of biological difference, but something that can and should be challenged and changed. However, as to agreement as to what causes this disadvantaged situation there are competing views, therefore feminism is not a unified ideology. Many contemporary feminists prefer to use the plural 'feminisms' to reflect the different perspectives that must be given a voice and consideration (McLeish, 1993). Feminism can also be interpreted as both a socio-political theory and a social movement. The most commonly recognised types of feminism are shown in Box 8.5.

Although feminism may be viewed as a relatively new concept, it origins can be traced back to at least the late-eighteenth century, a time when early capitalist development and the Industrial Revolution involved a marked change for the worse in the legal and economic situation of many women. In the late-eighteenth century women had no vote; were debarred from many occupations; had no legal standing beyond that of children; and married women had no legal property of their own, nor had any real right to divorce, even if abused by their husband.

A notable landmark in terms of the advocating of rights for women was Mary Wollstonecraft's *Vindication of the Rights of Woman*, published in 1792 partly in response to Tom Paine's *Rights of Man* published in 1791, which was written during the early years of the French Revolution. Wollstonecraft's perspective on feminism is one of a liberal persuasion, based upon a perspective that female subordination is rooted in a set of cultural and legal constraints that block women's opportunities and success in the public world. Wollstonecraft's demand for women's rights included the right to employment, financial independence, equal pay, childcare and reproductive rights. She attacked the belief held in society that women were intellectually and physically less capable then men (Tong, 1989). The significance of Wollstonecraft's work extended to later periods of feminism, influencing twentieth-century feminists such as Vera Brittain and Virginia Woolf, who campaigned on her ideas.

Box 8.5 Different types of feminism

Liberal feminism   Women are rational beings like men and therefore should have the same legal and political rights and the opportunity to compete equally with men in politics and paid employment.

Marxist feminism   Argues that women's oppression is a consequence of the class system and capitalism and equality can only be achieved when it is replaced by socialism. It also argues that the capitalist system was upheld by women's subordination. Similar to Engels, it holds the view that women's oppression originated with the introduction of private property, thereby removing any existing equality of the human community. The private ownership of the means of production, which is predominantly male, emphasises gender inequality.

Radical feminism   Argues that the patriarchal domination of women by men is the most basic form of power in society, and one that has its source in apparently private areas of life, notably in family and sexual relationships. The oppression of women through patriarchy is the most universal and fundamental form of domination and radical feminists aim to end this. It thus argues that neither liberal nor Marxist feminism is radical enough. The patriarchal system that oppresses and marginalises women cannot be reformed but must be dismantled. This would include not only the removal of all patriarchal legal and political structures, but also society's social and cultural institutions, especially the family and religions.

Modern socialist feminism   Seeks to combine the radical perspective with Marxist class analysis by exploring the relationship between capitalism and patriarchy. It rejects the view of Marxist patriarchs that women's oppression is not as important as workers' oppression.

Existentialist feminism   This view of feminism is characterised by Simone de Beauvoir's *The Second Sex*, which is probably the key thesis of twentieth-century feminism. The key theme of Beauvoir's work is that through defining the woman as the 'other' because she is not a man, she must make herself whatever she wants to be.

Sources: McLeish (1993); Tong (1989)

At the same time of Wollstonecraft, taking their inspiration from Revolutionary ideals, several women's clubs were formed in Paris and other cities to promote women's rights; the clubs being later dissolved by government decree. After the French Revolution it was in the US that feminism developed most quickly, becoming a model for the women's movement in other countries. American feminists were closely involved with groups committed to the abolition of slavery and to temperance (McLeish, 1993). However, the most widely cited point from which to date the feminist movement was the Women's Rights Convention held at Seneca Falls in America on 19 July 1848, at which principles of equal rights found a concrete expression (Bryson, 1993; Butalia, 1999).

However, feminism was not a movement or ideology exclusive to the West. Butalia (1999) draws attention to the roles of feminists both in India and Egypt at the beginning of the twentieth century. In Egypt, feminist Huda Saharawi opened the first school for girls in Egypt in 1910, and in 1919 organised women to demonstrate against British colonialism. In 1923 she founded the Egyptian Feminist Union, and a year later after attending an international conference for women in Italy, during the return trip she symbolically threw her veil into the Mediterranean, never to wear it again (Butalia, 1999). Attempts at suffrage, meaning a demand for the right for women to vote, were also made in Iran in 1911 and 1920, the Philippines in 1907, China in 1911, India in 1917, Japan in 1924 and Sri Lanka in 1927. Women not in the West were also waging another kind of battle; against colonialism, and for national independence.

A major figure of the feminist movement at the beginning of the twentieth century was Emmeline Pankhurst. She was a leader of the suffragette movement, campaigning on the slogan 'Votes for Women: Chastity for Men'. The 'suffragettes' was the name given to the militant wing of the suffrage campaign, which was demanding the right to vote for women. The suffrage movement was founded in 1866, when the British government ignored a petition signed by 1,500 women demanding full voting rights for women. In response the organisers established the National Society for Women's Suffrage. The frustration in the lack of progress that the non-violent suffragettes were making led Emmeline Pankhurst and her daughters Christabel and Sylvia to establish the more militant 'Women's Social and Political Union' in 1903 (Kelbie and Bloomfield, 2004). Typical militant campaigns included campaigns of arson, window smashing, and general public disruption, with the extent of civil disorder leading to more than 1,000 suffragettes receiving criminal records between 1903 and 1914 (Kelbie and Bloomfield, ibid.).

However, despite the eventual success of gaining the right to vote in many societies, legislation for more equal rights and a louder political voice, women still remain marginalised and disadvantaged to varying degrees in societies across the world. Thus the issues of feminism remain as globally relevant as they were more than 200 years ago. Although the term 'post-feminism' is sometimes used to suggest that either the agenda of feminism has been completed or has failed and is no longer relevant, feminists would reject the use of this term, arguing that it has been imposed from patriarchal sources. Issues of employment – access to it, equal rights within, and empowerment through it – remain critical issues for women's equality. One source of opportunity for possible employment for women in an increasing number of regions in the world is tourism.

Feminist studies and tourism

Given the impacts of tourism upon women, and the potential opportunities it offers, it is perhaps surprising that the application of feminist studies to tourism is not more progressed, especially when considered in relation to the abundant literature on the subject of gender and development (Richter, 1995; Ateljevic, 2004). Following the feminist perspective of society, the focus of the feminist enquiry of tourism is upon how the power relationships that exist between men and women manifest themselves within the context of tourism. As Kinnaird *et al.* (1994) point out, tourism emerges from gendered societies and subsequently aspects of tourism development exemplify those relations. This remit would include the representation of women; the employment opportunities; how women who are residents in tourism destinations are affected by tourism development; and the experiences of women as travellers. It is also important to realise that women are not just impacted upon by tourism but will also respond to opportunities for tourism.

However, many of these themes that embody issues of women's empowerment and disempowerment, and how tourism has influenced their domestic and social roles, are under-researched. Questioning the focus of existing research on the gender issues in tourism, Scheyvens (2002: 171) comments: 'One could almost be forgiven for assuming that gender issues in tourism are mainly concerned with sex tourism, such has been the focus on this subject in recent years.' A key issue of sex tourism is how women are presented and the embodied power relationships with their representation, as is highlighted in Box 8.6.

Although the example described in Box 8.6 could hardly be labelled implicit, there are many implicit sexual messages conveyed in tourism advertising that encourage sexual relationships as part of the tourism experience. While the possibility of developing sexual relationships may be an important motivator for tourism for both males and females, how they are presented is often male-dominated, as Kinnaird and Hall (1994: 11) comment: 'Signs and symbols, maths and fantasies that the tourism industry use to market destinations are often male orientated.'

The use of women as sexual objects to promote tourism destinations may have implications for women not only in tourism destinations but also within the societies that

---

Box 8.6 'Discover weapons of mass distraction'

This was the theme of an advertisement used by a low-cost airline in Britain to promote its flights during the war following the invasion of Iraq by American and British troops, on the premise of Saddam Hussein having 'weapons of mass destruction'. Underneath the title was a picture of a woman with enormous breasts and the names of destinations you could fly to. Posters of the advertisement appeared on billboards and at bus stops. It was the second most complained-about advertisement to the advertising standards authority (ASA) in Britain in 2003. As Redfern (2003: 6) comments: 'Some felt the ad trivialised the Iraq War, and was insensitive and tasteless. Some felt its implied message (Go on holiday – ogle breasts) was irresponsible and encouraged sex tourism.'

Think point

What do you think about the advertisement described in Box 8.6? In your view is this advertisement 'sexist'?

---

tourists originate from. Returning to the concept of reciprocity that is inherent to the tourism system, and Urry's (1990) perspective that tourism has a much longer-term impact on our knowledge and perception of the world than short-term periods of travel may imply, it is conceivable that the experiences of sex tourism may also affect gender relations in tourism-generating areas. Subsequently, not only may women be viewed by males as sex objects in tourism destinations but also increasingly in their home societies. In trying to explain in which destination sex tourism is most likely to develop, Kinnaird and Hall (1994) comment that it is more likely to flourish in societies that have particular gender and power relationships and find themselves interacting with modernising and globalising influences.

The consequences for a destination society that gains a reputation for sex tourism is that its visitors are likely to be skewed towards men and single visitors. In Thailand, a country that has developed such a reputation based upon its sex clubs, strip clubs and prostitution, especially in Bangkok and Pattaya, in 1999 only 35 per cent of visitors to Thailand were women (Bowes, 2004). The Thai Tourist Authority were concerned about this imbalance, and also that the sleazy image of a small part of Thailand should dominate the country's image over its range of environmental and cultural attractions. Apart from helping to provide a healthier balance in terms of the cultural and social impacts of tourism, attracting women travellers is financially important because the increased economic and personal independence of women in western countries makes them an increasingly powerful market segment. Thus the Thai Tourist Authority now actively promotes the safety of its Buddhist culture and its spa, health, shopping and cooking holidays in a bid to win this market. In 2004, the percentage of the total numbers of visitors to Thailand who were female had risen to 40 per cent (Bowes, ibid.).

In terms of the other main area of tourism enquiry within the field of gender and tourism, that of employment, tourism can potentially offer women the opportunity for economic autonomy and greater influence in household decision-making (Scheyvens, 2002). She also points out that far from being victims of tourism development, women work in innovative ways to secure benefits for themselves and for their families. Potentially, tourism may offer women opportunities for empowerment, not just economically, but also psychologically and socially. Thus, besides leading to financial independence for women, employment in tourism may increase confidence and self-esteem, and also lead to women having a more influential social and perhaps political role in society. However, as Kinnaird *et al.* (1994) observe, tourism operates with the wider economic, cultural and political contexts, all of which will influence how opportunities for women will manifest themselves.

Consequently, Sinclair (1997) suggests that the existing empirical research points to a confusing image of the extent to which employment in tourism may or may not offer empowerment for women. Scheyvens (2002) observes that, to date, the literature of women's experiences of employment in the tourism industry is overwhelmingly negative with women occupying the majority of low-skilled and low-waged jobs. How women are stereotyped within cultures will also influence the employment opportunities for women in the tourism industry as is described in Box 8.7.

Thus, even though in many cases tourism has opened up more employment opportunities for women, and globally women occupy over half the positions available in the tourism industry, there is criticism that tourism reinforces the gendered division of labour (Scheyvens, 2002). Women are also often expected to fulfil domestic and social responsibilities besides carrying out employment. This can subsequently lead to domestic tension if women no longer have time to carry out their defined and expected domestic roles or dare to challenge them. The economic independence of women may also challenge the role of the male as the main wage-earner in the family and decision-maker over how money is allocated.

Conversely, it can be argued that some types of tourism employment offer attractive opportunities for women because of their time flexibility and demand for cultural norms of hospitality, which makes them convenient to fit around family responsibilities. For example, in many island communities throughout the Mediterranean region, it is often women who are the first to seize the financial opportunities of tourism as an activity where their traditional areas of competence can be used (Scott, 2001). Also, in Barbados, women beach-vendors are generally older than their male counterparts, and work because the job offers the flexibility of time to combine it with domestic duties (Momsem, 1994). However, such roles of employment do little to challenge the existing inequalities between men and women in society, which lies at the core of the feminist argument.

The type of tourism development may also be influential in determining the extent to which employment challenges existing stereotypes and unequal gendered power relationships. For example, Retour, a Dutch NGO, have undertaken work with the Maasai

---

Box 8.7 Cultural stereotyping of women

The cases of the Caribbean, the Philippines and Indonesia illustrate how the stereotyping of women within cultures influences their employment opportunities. For example, in the Caribbean the main form of employment for women in the tourism industry is as maids in hotels, which can be viewed as an extension of their domestic role and subsequently can be classified as unskilled and paid accordingly (Momsem, 1994). Although the development of training institutions for the tourist industry has helped some women move into better positions as receptionists and housekeepers, men are more likely to have opportunities to take up training places. The opportunities for women both to train and to be employed in tourism, are often restricted by expected family and other duties (Momsem, ibid.). Another example of how the cultural positioning of women also influences their likely employment opportunities is in the Philippines and Indonesia, where they are more likely to receive front-line jobs, as they are considered to be more physically attractive to tourists and servile than men.

Think point

Summarise the advantages and disadvantages of tourism as a form of employment for women.

---

in northeast Tanzania, to encourage the empowerment of women through the development of human and environmentally sensitive tourism. The role and status of women in Maasai society is determined by their relationship to men, i.e. as a mother, daughter or wife (Retour, 2004). Women are not allowed to have possessions of their own and have no income. Through working with men in the community, Retour staff eventually persuaded them that women should become actively involved in tourism, for two main reasons. First, development agency funding would not otherwise be forthcoming, and second, the 'ethical' types of tourists they wanted to attract would be unlikely to be interested in a society that oppressed women (Schevyens, 2002). The women now work in providing campsites, walking safaris and beadwork workshops, retaining the income they gain through selling beadwork.

A further example of how women have gained through the development of alternative tourism is in the Dadia region of the north of Greece. The development of a women's cooperative, to run a restaurant and supply local goods to a shop in an ecotourism development based upon the conservation of raptor species, has led to the economic and psychological empowerment of women (Svoronou and Holden, 2004). However, care needs to be taken in making broad assumptions that alternative tourism is more likely to lead to women's empowerment than mainstream tourism development. As has been pointed out, tourism does not exist in isolation of the wider political, economic, cultural and social influences that determine women's roles in society. Yet, it is suggested that where tourism development depends on donor money, for example from the United Nations, and is established with a target market of ethical or more responsible tourists, there would seem to be greater opportunities for the emancipation of women.

The issues of women's roles in tourism are therefore complex. They need research both to enhance the opportunities for women to work and fulfil their potential in the industry, and also to understand the effects upon their families and societies. Similar to ethics, the application of feminist studies to tourism is an underdeveloped area. Yet both are important routes of enquiry, given the changes that tourism can potentially induce in societies and nature, and are likely to become more important in the future as more emphasis is placed upon equality and human rights as part of the development process.

■ ■ ■ ■ ■ ■

Summary

- **Ethics may be thought of as the branch of philosophy that deals with moral problems and judgements. The question of where morality has come from is one that has been of interest since at least the times of the Ancient Greeks. Some philosophers such as Plato were 'moral absolutists', believing that objective moral truths and facts existed. A contrasting position to moral absolutism is that** of 'moral relativism', which takes the view that the beliefs and values of different cultures are varied but ethically valid.

■ Other major theories of morality include 'egoism'; 'utilitarianism'; and the 'categorical imperative'. However, attempting to understand the morality of what is 'right' or 'wrong' and rules to base our actions upon is a complex and difficult undertaking. Consequently, evaluating the 'right' or 'wrong' of tourism is difficult.

■ Tourism raises many ethical questions about how groups of people and individuals interact with each other within the tourism system. Many of these issues are related to a wider social context, for example whether it is 'right' or 'wrong' to visit countries and subsequently economically support governments who deny basic human and democratic rights. However, the ethical questioning within tourism extends beyond the boundaries of how we interact with each other, to also include how we interact with our natural surroundings. In this context it is also necessary to consider theories from the field of environmental ethics.

■ Embodied within the tourism system are power relationships between men and women. An obvious ethical issue of this relationship is sex tourism. However, more typically tourism acts as a catalyst to changes in women's roles in society, which may bring confrontation with expected norms. Important in the analysis of these changes is feminist studies.

■ Following the feminist perspective of society, the focus of the feminist enquiry of tourism is based upon how the power relationships that exist between men and women manifest themselves within the context of tourism. Feminists share the view that patriarchy is not natural but something that can and should be challenged. However, there are contrasting views of the causes of inequality, hence many contemporary feminists refer to 'feminisms'. The remit of the typical issues of the feminist enquiry of tourism would include the representation of women; employment opportunities; how women who are residents in tourism destinations are affected by tourism development; and the experiences of women as travellers.

■ Employment in tourism potentially offers women economic, psychological and social empowerment. However, tourism operates within the wider economic, cultural and political contexts of society, all of which will influence how opportunities for women will manifest themselves. To date, the literature of women's experiences of employment in the tourism industry is overwhelmingly negative, with women occupying the majority of low-skilled and low-waged jobs.

Suggested reading

Jenkins, T. (ed.) (2002) 'Ethical Tourism: Who Benefits?', Hodder and Stoughton, London.
Robinson, D. and Garratt, C. (1999) Introducing Ethics, Icon Books, Cambridge.
Sinclair, M.T. (1997) Gender, Work and Tourism, Routledge, London.
Smith, M. and Duffy, R. (2003) The Ethics of Tourism Development, Routledge, London.
Tong, R. (1989) Feminist Thought: A Comprehensive Introduction, Routledge, London.
WTO (1999) Global Code of Ethics for Tourism, World Tourism Organisation, Madrid.

CONCLUDING
NOTE

It is not the intention of this concluding note to reiterate the main points of the book, which have been summarised at the end of each chapter. However, having worked their way through the preceding chapters, the reader will now hopefully be more aware of the complexity of tourism, and had their preconceived ideas of it expanded and challenged. A major theme of this book is that just as the social sciences emerged from changes in society, so have wider economic and social forces shaped tourism. In terms of searching for a critical juncture in the evolution of society that created the conditions for a mass participation in tourism, the key event was the Industrial Revolution. The political, economic, social and technological changes that this revolution brought were instrumental in shaping patterns of contemporary tourism.

At the beginning of the twenty-first century, the importance of tourism in many societies around the world is demonstrated by an emerging recognition of it as a part of social citizenship. It has also become for many an essential part of their lifestyle, and also a means to formulate identity at a time when social class is becoming less relevant as a determinant of social behaviour. Yet it is also important to remember that the majority of the world's population are denied the opportunity to participate in and enjoy the benefits of tourism, most usually as a consequence of poverty.

The system of tourism can subsequently not be separated from the wider economic, political and social structures that exist in society. It encompasses a variety of stakeholders and consequently a range of economic and power relationships within and between countries. These influence not only who becomes a tourist but also who are the recipients of the benefits and costs of tourism. Consequently, the structures of the wider political economy influence how the resources and benefits of tourism will be distributed both globally and locally.

The role and use of tourism in society is multi-faceted and constantly demonstrates duality. For example, it can be: a means of economic enhancement and used to fight poverty but may also reiterate economic inequality; a destroyer of culture or a source for renewed cultural pride; a destructive force upon nature or a means for environmental conservation; a representation of the escape from the everyday pressures of life or a search for authenticity; an activity that offers empowerment for women or that exploits women in a patriarchal and stereotypical fashion. Which of these aspects of tourism manifests themselves will ultimately be a consequence and reflection of the wider relationships that exist in society, and the ethical base upon which they are formulated.

The impacts that result from the interactions that take place within the tourism system potentially raise a number of ethical issues. These include democracy, denial of human rights, the destruction of nature, exploitation and abuse. Given the global scale and the increasing economic importance of tourism, debates about the ethics of tourism are likely to become more prevalent in the future. Therefore the requirement for a deeper understanding of tourism and its role in society has never been more relevant than today.

BIBLIOGRAPHY

Abercrombie, N., Hill, S. and Turner, B.S. (2000) *The Penguin Dictionary of Sociology*, 4th edn, Penguin, London.

Argyle, M. (1994) *Social Psychology of Leisure*, Penguin, London.

Ali, N. (2004) Unpublished PhD research, University of Luton, Luton.

Ashley, C., Roe, D. and Goodwin, H. (2001) *Pro-Poor Tourism Strategies: Making Tourism Work for the Poor: A Review of Experience*, Report No. 1, Overseas Development Institute, London.

Ateljevic, I. (2004) *Gender Researchers in Tourism Studies Network*, email correspondence, trinet-l@hawaii.edu.

Atkinson, L.R., Atkinson, C.R. and Hilgard, R.E. (1983) *Introduction to Psychology*, 8th edn, Harcourt Brace Jovanich, New York.

Babbie, E. (1995) *The Practice of Social Research*, 7th edn, Wadsworth Publishing Company, Belmont.

Badger, A., Barnett, P., Corbyn, L. and Keefe, J. (1996) *Trading Places: Tourism as Trade*, Tourism Concern, London.

Barke, M. and France, L.A. (1996) 'The Costa del Sol' in Barke, M., Towner, J. and Newton, M.T. (eds) *Tourism in Spain: Critical Issues*, CAB International, Wallingford, 265–308.

Barker, L.M. (1982) 'Traditional landscape and mass tourism in the Alps', *Geographical Review*, 72(4): 395–415.

Baudrillard, J. (1983) *Fatal Strategies*, Pluto Press, London.

Baumol, W.J. and Blinder, A.S. (1999) *Economics: Principles and Policy*, 8th edn, The Dryden Press, Fort Worth.

Becheri, E. (1991) 'Rimini and Co. – the end of a legend?: dealing with the algae effect', *Tourism Management*, 12(3): 229–235.

Begg, D., Fischer, S. and Dornbusch, R. (2003) *Economics*, 7th edn, McGraw-Hill, London.

Benjamin, A. (2004) 'Countryside Retreat', *The Guardian*, London, 28 January, http://society.guardian.co.uk.

Bianchi, R. (1999) *A Critical Ethnography of Tourism Entrepreneurship and Social Change in a Fishing Community in Gran Canaria*, unpublished PhD thesis, University of North London.

Bianchi, R. (2002) 'Towards a new political economy of global tourism' in Sharpley, R. and Telfer, D. (eds) *Tourism and Development: Concepts and Issues*, Channel View Publications, Clevedon, 265–299.

Bird, B. (1989) *Langkawi: From Mahusri to Mahathir: Tourism for Whom?*, INSAN, Kuala Lumpur.

Blue Flag (2004) http://www.blueflag.org

Bocock, R. (1993) *Consumption*, Routledge, London.

Boissevain, J. (ed.) (1996) *Coping with Tourists: European Reactions to Mass Tourism*, Berghahn Books, Oxford.

Boo, E. (1990) *Ecotourism: The Potentials and Pitfalls*, Vol. 1, World Wide Fund for Nature, Washington.

Boorstin, D.J. (1961) *The Image: A Guide to Pseudo-events in America*, reprinted (1992) First Vintage Books, New York.

Boulding, K.E. (1973) 'The economics of the coming spaceship earth' in Daly, H.E. (ed.) *Toward a Steady-state Economy*, W.H. Freeman and Company, San Francisco.

Bowcott, O., Traynor, I., Webster, P. and Walker, D. (1999) 'Analysis: green politics', *The Guardian*, London, 11 March: 17.

Bowes, G. (2004) 'Thais target women to shed sleazy image', *The Observer*, London, 17 October: 10.

Bramwell, B. and Lane, B. (1993) 'Interpretation and sustainable tourism: the potential and the pitfalls', *Journal of Sustainable Tourism*, 1(2): 71–80.

Bray, R. and Raitz, V. (2001) *Flight to the Sun: The Story of the Holiday Revolution*, Continuum, London.

Brendon, P. (1991) *Thomas Cook: 150 Years of Popular Tourism*, Secker and Warburg, London.

Briguglio, L. and Briguglio, M. (1996) 'Sustainable tourism in the Maltese Isles' in Briguglio, L., Butler, R., Harrison, D. and Filho, W.L. (eds) *Sustainable Tourism in Islands and Small States*, Pinter, London, 161–179.

Britton, S.G. (1982) 'The political economy of tourism in the Third World', *Annals of Tourism Research*, 9: 331–358.

Brohman, J. (1996) 'New directions in tourism for Third World development', *Annals of Tourism Research*, 23(1): 48–70.

Brown, F. (1998) *Tourism Reassessed: Blight or Blessing?*, Butterworth-Heinemann, Oxford.

Bryson, V. (1993) 'Feminism' in Eatwell, R. and Wright, A. (eds) *Contemporary Political Ideologies*, Pinter, London, 192–215.

Buckingham, L. and Wolf, J. (1999) 'EC to look at Airtours bid', *The Guardian*, London, 4 June: 10.

Bull, A. (1991) *The Economics of Travel and Tourism*, Pitman, London.

Burac, M. (1996) 'Tourism and environment in Guadeloupe and Martinique' in Briguglio, L., Butler, R., Harrison, D. and Filho, W.L. (eds) *Sustainable Tourism in Islands and Small States: Case Studies*, Pinter, London, 63–74.

Burchell, J. and Lightfoot, S. (2001) *The Greening of the European Union?: Examining the EU's Environmental Credentials*, Continuum, London.

Burke, P. (1978) *Popular Culture in Early Modern Europe*, Wildwood House, Aldershot.

Burningham, D., Bennett, P., Cave, M., Herbert, D. and Higham, D. (1984) *Teach Yourself Economics*, Hodder and Stoughton, Sevenoaks.

Burns, P. (1993) 'Sustaining tourism employment', *Journal of Sustainable Tourism*, 1(2): 81–96.

Burns, P. (1999) *An Introduction to Tourism and Anthropology*, Routledge, London.

Burns, P. and Holden, A. (1995) *Tourism: A New Perspective*, Prentice-Hall, Hitchin.

Burrell, G. and Morgan, G. (1979) *Sociological Paradigms and Organisational Analysis*, Heinemann, London.

Burrell, R. (1989) *The Greeks*, Oxford University Press, Oxford.

Butalia, U. (1999) 'Domestic murder and the golden sea', *The New Internationalist*, 309: 18–20.

Butcher, J. (2002) 'Weighed down by ethical baggage' in Jenkins, T. (ed.) *Ethical Tourism: Who Benefits?*, Hodder and Stoughton, London, 59–74.

Butler, R. (1980) 'The concept of a tourist area cycle of evolution: implications for management of resources', *The Canadian Geographer*, 24: 1–12.

Butler, R. (1998) 'Sustainable tourism – looking backwards in order to progress' in Hall, M.C. and Lew, A.A. (eds) *Sustainable Tourism: A Geographical Perspective*, Longman, Harlow, 25–34.

Carr, E.H. (1990) *What is History?*, Penguin, London.

Castree, N. (2003) 'Place: connections and boundaries in an interdependent world' in Holloway, S.L., Rice, S.P. and Valentine, G. (eds) *Key Concepts in Geography*, Sage, London, 165–186.

Cater, E. (1992) 'Profits from paradise', *Geographical Magazine*, 64(3): 17–20.

Cater, E. (1993) 'Ecotourism in the Third World: problems for sustainable tourism development', *Tourism Management*, April: 85–90.

Cater, E. (1994) 'Ecotourism in the Third World – problems and prospects for sustainability' in Cater, E. and Lowman, G. (eds) *Ecotourism: A Sustainable Option*, John Wiley & Sons, Chichester, 3.

Chaudhuri, A. (1999) 'Unwelcome guests', *The Guardian*, London, 20 October, http://education.guardian.co.uk.

Clark, A. (2004) 'How low can they go?', *The Guardian*, London, 5 June: 12.

Clarke, J. and Critcher, C. (1985) *The Devil Makes Work: Leisure in Capitalist Britain*, Macmillan Press, Basingstoke.

Cleaver, T. (2002) *Understanding the World Economy*, 2nd edn, Routledge, London.

Club Freestyle (1999) *Summer '99: Have it Your Way*, 2nd edn, Thomson Holidays, London.

Coccossis, H. and Parpairis, A. (1996) 'Tourism and carrying capacity in coastal areas: Mykonos, Greece' in Priestley, G.K., Edwards, J.A. and Coccossis, H. (eds) *Sustainable Tourism: European Experiences*, CAB International, Wallingford, 153–175.

Coghlan, T. (2004) 'Bin Laden's hideout is touted as a tourist site', *The Daily Telegraph*, London, 4 December: 17.

Cohen, E. (1979) 'A phenomenology of tourist experiences', *Sociology*, 13: 179–201.

Cohen, E. (1984) 'The sociology of tourism: approaches, issues and findings', *Annual Review in Anthropology*, 10: 373–392.

Collins (2004) *Geography Dictionary*, HarperCollins, Glasgow.

Cooper, C., Fletcher, J., Gilbert, D., Wanhill, S. and Shepherd, R. (1998) *Tourism: Principles and Practices*, 2nd edn, Longman, Harlow.

Cooper, D.E. (2001) 'Arne Naess' in Palmer, J.A. (ed.) *Fifty Key Thinkers on the Environment*, Routledge, London, 211–216.

Crandall, R. (1980) 'Motivations for leisure', *Journal of Leisure Research*, 12: 45–54.

Crang, M. (1999) 'Knowing tourism and practices of vision' in Crouch, D. (ed.) *Leisure/Tourism Geographies: Practices and Geographical Knowledge*, Routledge, London.

Crick, M. (1988) 'Sun, sex, sights, savings and servility: representations of international tourism in the social sciences', *Criticism, Heresy and Interpretation (CHI)*, 1(1): 37–76; republished (1989) in *Annual Review of Anthropology*, 18: 307–344.

Crompton, J. (1979) 'Why people go on a pleasure vacation', *Annals of Tourism Research*, 6(4): 408–424.

Crouch, D. (ed.) (1999) *Leisure/Tourism Geographies: Practices and Geographical Knowledge*, Routledge, London.

Crystal, D. (ed.) (1994) *The Cambridge Encyclopaedia*, 2nd edn, Cambridge University Press, Cambridge.

Csikszentmihalyi, M. (1975) *Beyond Boredom and Anxiety*, Jossey-Bass, San Francisco.

Csikszentmihalyi, M. (1997) *Living Well*, Weidenfeld & Nicolson, London.

Dalton, R.J. (1993) 'The environmental movement in western Europe' in Kamieniecki, S. (ed.) *Environmental Politics in the International Arena: Movements, Parties, Organisations and Policy*, State University of New York Press, New York.

Dann, G. (1977) 'Anomie, ego-enhancement and tourism', *Annals of Tourism Research*, 4(4): 184–194.

Dann, G. (1981) 'Tourism motivation: an appraisal', *Annals of Tourism Research*, 8(2): 187–219.

Dann, G. and Cohen, E. (1991) 'Sociology and tourism', *Annals of Tourism Research*, 18: 155–169.

D'Auvergne, B.D.E. (1910) *Switzerland in Sunshine and Snow*, T. Wener Laurie, London.

Davidoff, L.L. (1987) *Introduction to Psychology*, 3rd edn, McGraw-Hill Book Company, New York.

Davidoff, L.L. (1994) *Introduction to Psychology*, 4th edn, McGraw-Hill Book Company, New York.

Davidson, J. and Spearritt, P. (2000) *Holiday Business: Tourism in Australia Since 1870*, Melbourne University Press, Victoria.

Decrop, A. (1999) 'Tourists' decision-making and behaviour processes' in Pizam, A. and Mansfield, Y. (eds) *Consumer Behaviour in Travel and Tourism*, The Haworth Hospitality Press, Binghamton, 103–133.

Dibb, S., Simkin, L., Pride, W.M. and Ferrell, O.C. (1994) *Marketing: Concepts and Strategies*, 2nd edn, Houghton Mifflin, Boston.

Dicks, B. (2003) *Culture on Display: The Production of Contemporary Visitability*, Open University Press, Maidenhead.

Dieke, P.U.C. (ed.) (2000) *The Political Economy of Tourism Development in Africa*, Cognizant Communication Corporation, New York.

Doxey, G. (1975) 'A causation theory of visitor–resident irritants: methodology and research inferences' in *Proceedings of the Travel Research Association*, 6th Annual Conference, San Diego.

Doyle, T. and McEachern, D. (1998) *Environment and Politics*, Routledge, London.

Drumm, A. (1995) *Converting from Nature Tourism to Ecotourism in the Ecuadorian Amazon*, Paper given at the World Conference on Sustainable Tourism, Lanzarote, April.

Dudenhoefer, D. (2002) 'Selling nature without destroying it in Bolivia', *Choices*, New York.

Dumazedier, J. (1967) *Towards a Society of Leisure*, Free Press, New York.

Eadington, W.R. and Smith, V.L. (1992) 'The emergence of alternative forms of tourism' in Smith, V. L. and Eadington W.R. (eds) *Tourism Alternatives: Potentials and Problems in the Development of Tourism*, University of Pennsylvania Press, Philadelphia.

Edington, J. and Edington, A. (1986) *Ecology, Recreation and Tourism*, Cambridge University Press, Cambridge.

EIU (1992) *The Leakage of Foreign Exchange Earning from Tourism*, Economist Intelligence Unit, London.

Elliott, J.A. (1994) *An Introduction to Sustainable Development: The Developing World*, Routledge, London.

Ellwood, W. (2001) *No-nonsense Guide to Globalization*, New Internationalist Publications, Oxford.

Elyne, M. (1942) *Australia's Alps*, Angus & Robertson, Sydney.

Engels, F. (1845) *The Condition of the Working Class in England*, republished (1987) by Penguin, London.

European Commission Directorate General (1998) *Facts and Figures on the Europeans on Holidays*, Eurostat, Brussels.

European Tourism Analysis (1993) *Ten Main Characteristics for Quality Tourism*, BAT-Leisure Research Institute, Hamburg.

Evans, G. (1993) 'Tourists rush for kill a seal pup holiday', *Evening Standard*, London, 5 July.

Ezard, J. (1998) 'Ship ahoy', *The Guardian*, London, 23 March: 8.

Fearis, B. (2004) 'Paint the world pink', *The Observer*, London 29 August, http://observer.guardian.co.uk.

Fennell, D. (1999) *Ecotourism: An Introduction*, Routledge, London.

Fennell, D. (2000) 'Tourism and applied ethics', *Tourism Recreation Research*, 25(1): 59–69.

Fielding, K.A., Pearce, P.L. and Hughes, K. (1995) 'Case study, climbing Ayres Rock: relating visitor motivation, time perception and enjoyment' in McIntosh, R.W., Goeldner, C.R. and Brent-Ritchie, J.R. (eds) *Tourism: Principles, Practices, Philosophies*, 7th edn, John Wiley & Sons, New York, 179–190.

Finnish Tourist Board (1993) *Key Goals and Strategies for 1993–95*, Finnish Tourist Board, Helsinki.

Fodness, D. (1994) 'Measuring tourist motivation', *Annals of Tourism Research*, 21(3): 555–581.

Frank, A.G. (1967) *Capitalism and Underdevelopment in Latin America: Historical Studies of Chile and Brazil*, Monthly Review Press, New York.

Franklin, A. (2003) *Tourism: An Introduction*, Sage Publications, London.

Friends of the Earth (1997) *Atmosphere and Transport Campaign*, www.foe.co.uk.

Giddens, A. (1999) *Runaway World: How Globalisation is Reshaping Our Lives*, Profile Books, London.

Giddens, A. (2001) *Sociology*, 4th edn, Polity, Cambridge.

Gill, R. (1967) *Evaluation of Modern Economics*, Prentice Hall, Upper Saddle River.

Gningue, A.M. (1993) 'Integrated rural tourism Lower Casamance' in Eber, S. (ed.) *Beyond the Green Horizon: A Discussion Paper on the Principles for Sustainable Tourism*, World Wide Fund for Nature, Godalming.

Goffman, I. (1959) *The Presentation of Self in Everyday Life*, Penguin, Harmondsworth.

Goldsmith, E. (1974) 'Pollution by tourism', *The Ecologist*, 4(2): 9–10.

Good, R. and Grenier, P. (1994) 'Some environmental impacts of recreation in the Australian Alps', *Australian Parks and Recreation*, 30(4): 20–26.

Gottileb, A. (2000) in Monk, R. and Raphael, F. (eds) *The Great Philosophers*, Phoenix, London, 5–46.

Goudie, A. and Viles, H. (1997) *The Earth Transformed: An Introduction to Human Impacts on the Environment*, Blackwell, Oxford.

Graburn, N.H.H. (1989) 'Tourism: the sacred journey' in Smith, V.L. (ed.) *Hosts and Guests: The Anthropology of Tourism*, 2nd edn, University of Pennsylvania, Philadelphia.

Graburn, N.H.H. (2001) 'Secular ritual: a general theory of tourism' in Smith,V.L. and Brent, M. (eds) *Hosts and Guests Revisited: Tourism Issues of the 21st Century*, Cognizant Communications Corporation, New York, 42–50.

Graburn, N.H.H. and Jafari, J. (1991) 'Introduction: tourism and the social sciences', *Annals of Tourism Research*, 18(1): 1–11.

Gross, D.R. (1992) *Psychology: The Science of Mind and Behaviour*, 2nd edn, Hodder and Stoughton, London.

Gunn, C. (1994) *Tourism Planning: Basic, Concepts, Issues*, 2nd edn, Taylor & Francis, Washington.

Hall, C. and Lew, A. (eds) (1998) *Sustainable Tourism: A Geographical Perspective*, Addison Wesley Longman, Harlow.

Hall, C.M. (1994) *Tourism and Politics: Policy, Power and Place*, John Wiley & Sons, Chichester.

Hall, C.M. and Page, S.J. (1999) *The Geography of Tourism and Recreation: Environment, Place and Space*, Routledge, London.

Haralambos, M. and Holborn, M. (1990) *Sociology: Themes and Perspectives*, 3rd edn, Unwin Hyman, London.

Hardin, G. (1968) 'The tragedy of the commons', *Science*, 162: 1243–1248.

Harrison, D. (1988) *The Sociology of Modernisation and Development*, Unwin Hyman, reprinted (1991) by Routledge, London.

Harrison, D. (ed.) (1992) *Tourism and the Less Developed Countries*, John Wiley & Sons, Chichester.

Harrison, D. (ed.) (2001) *Tourism and the Less Developed World: Issues and Case Studies*, CAB International, Wallingford.

Harvey, L., MacDonald, M. and Hill, J. (2000) *Theories and Methods*, Hodder and Stoughton, Abingdon.

Heilbroner, R. and Thurow, L. (1998) *Economics Explained*, Touchstone, New York.

Hitchcock, M. (2000) 'Introduction' in Hitchcock, M. and Teague, K. (eds) *Souvenirs: The Material Cultures of Tourism*, Ashgate, Aldershot, 1–18.

Hobsbawm, E. (1962) *The Age of Revolution*, Abacus, London.

Hobsbawm, E. (1975) *The Age of Capital*, Abacus, London.

Hodgson, G. (1996) *People's Century*, BBC Books, London.

Holden, A. (2000) *Environment and Tourism*, Routledge, London.

Hollinshead, K. (1990) 'The ethics of ennoblement: a review of Leopold on leisure and of Callicott on Leopold', *Society and Natural Resources*, 3: 373–383.

Holloway, C.J. (1998) *The Business of Tourism*, 5th edn, Longman, Harlow.

Holloway, S.L., Rice, S.P. and Valentine, G. (eds) (2003) *Key Concepts in Geography*, Sage, London.

Holmes Rolston III (1988) *Environmental Ethics: Duties to and Values in the Natural World*, Temple Press, Philadelphia.

Holt-Jensen, A. (1999) *Geography: History and Concepts: A Students' Guide*, 3rd edn, Sage, London.

House, J. (1997) 'Redefining sustainability: a structural approach to sustainable tourism' in Stabler, M. (ed.) *Tourism and Sustainability: Principles to Practice*, CAB International, Wallingford, 89–104.

Hudman, E. (1991) 'Tourism's role and response to environmental issues and potential future effects', *Revue de Tourisme (The Tourist Review)*, 4: 17–21.

Hughes, H. (1998) 'Sexuality, tourism and space: the case of gay visitors to Amsterdam' in Tyler, D., Guerrier, Y. and Robertson, M. (eds) *Managing Tourism in the Cities*, John Wiley & Sons, Chichester, 163–178.

Hughes, H. (2002) 'Gay men's holidays: identity and inhibitors' in Clift, S., Luongo, H. and Callister, C. (eds) *Gay Tourism: Culture, Identity and Sex*, Continuum, London, 174–190.

Hughes, J. (1990) *The Philosophy of Social Research*, 2nd edn, Longman, Harlow.

Hunter, C. and Green, H. (1995) *Tourism and the Environment: A Sustainable Relationship?*, Routledge, London.

Icelandic Tourist Board (2003) *Tourism and Iceland in Figures*, Icelandic Tourist Board, Reykjavik.

Inglis, F. (2000) *The Delicious History of the Holiday*, Routledge, London.

Iso-Ahola, E.S. (1980) *The Social Psychology of Leisure and Recreation*, Wm. C. Brown, Dubuque.

Iso-Ahola, E.S. (1982) 'Toward a social psychological theory of tourism motivation: a rejoinder', *Annals of Tourism Research*, 6: 257–264.

Ittleson, W.H., Franck, K.A., and O'Hanlon, T.J. (1976) 'The nature of environmental experience' in Wagner, S., Cohen, B.S., and Kaplan, B. (eds) *Experiencing the Environment*, Plenum Press, New York, 187–206.

Jafari, J. (1977) 'Editor's page', *Annals of Tourism Research*, 1: 1.

Jafari, J. (2001) 'The scientification of tourism' in Smith, V.L. and Brent, M. (eds) *Hosts and Guests Revisited: Tourism Issues of the 21st Century*, Cognizant Communications Corporation, New York, 28–41.

Jenkins, T. (ed.) (2002) *Ethical Tourism: Who Benefits?*, Hodder and Stoughton, London.

Jenner, P. and Smith, C. (1992) *The Tourism Industry and the Environment*, The Economist Intelligence Unit, London.

Johnston, E.M. (1992) 'Case study, facing the challenges: adventure in the mountains of New Zealand', in Wesley, B.S. and Hall, L. (eds) *Special Interest Tourism*, Belhaven, London, 159–169.

Johnston, R. (2003) 'Geography and the social science tradition' in Holloway, S.L., Rice, S.P. and Valentine, G. (eds) *Key Concepts in Geography*, Sage, London, 51–72.

Jordanova, L. (2000) *History in Practice*, Arnold, London.

Jowit, J. and Soldal, H. (2004) 'It's the new sport for tourists: killing baby seals', *The Observer*, London, 3 October: 3.

Keefe, J. (1995) 'Water fights', *Tourism in Focus*, 17: 8–9.

Kelbie, P. and Bloomfield, S. (2004) 'Ethel threw an egg at Churchill. After 90 years, it is time she was pardoned', *The Independent on Sunday*, London, 16 May: 14.

Kinnaird, V. and Hall, D. (eds) (1994) *Tourism: A Gender Analysis*, John Wiley & Sons, Chichester.

Kinnaird, V., Kothari, U. and Hall, D. (1994) 'Tourism: gender perspectives' in Kinnaird, V. and Hall, D. (eds) *Tourism: A Gender Analysis*, John Wiley & Sons, Chichester, 1–31.

Kirkby, S.J. (1996) 'Recreation and the quality of Spanish coastal waters', in Barke, M., Towner, J. and Newton, M.T. (eds) *Tourism in Spain: Critical Issues*, CAB International, Wallingford, 190–211.

Klemm, M. (2002) 'Tourism and ethnic minorities in Bradford: the invisible segment', *Journal of Tourism Research*, 41: 85–91.

Klemm, M. and Kelsey, S.J. (2002) *Catering for a Minority? Ethnic Groups and the British Travel Industry*, Management Paper, University of Bradford Management School.

Krippendorf, J. (1986) 'Tourism in the system of industrial society', *Annals of Tourism Research*, 13: 517–532.

Lanjouw, A. (1999) 'Mountain Gorilla Tourism in Central Africa', owner.mtn-forum @eige.apc.org

Laws, E. (1999) 'Package holiday pricing: cause of the IT industry's success or cause for concern', in Baum, T. and Mudambi, R. (eds) *Economic and Management Methods for Tourism and Hospitality Research*, John Wiley & Sons, Chichester, 197–214.

Lea, J.P. (1988) *Tourism and Development in the Third World*, Routledge, London.

Lea, J.P. (1993) 'Tourism development ethics in the Third World', *Annals of Tourism Research*, 20(4): 701–715.

Leach, E. (1996) *Levi-Strauss*, 4th edn, Fontane Press, London.

Lechte, J. (1994) *Fifty Key Contemporary Thinkers: From Structuralism to Postmodernity*, Routledge, London.

Leech, K. (2002) 'Enforced primitivism' in Fox C. (ed.) *Ethical Tourism: Who Benefits?*, Hodder and Stoughton, London, 75–94.

Leed, E.J. (1991) *The Mind of the Traveller: From Gilgamesh to Global Tourism*, HarperCollins, New York.

Leopold, A. (1949) *A Sand County Almanac*, Oxford University Press, Oxford.

Levitt, R. (2004) 'Glad to get away: the same sex guide to romantic holidays', *The Independent*, London, 8 February, http://travel.independent.co.uk

Lew, A. (1999) 'A place called tourism geographies', *Tourism Geographies*, 1(1): 1–2.

Lewis, R. and Wild, M. (1995) *French Ski Resorts and UK Ski Tour Operators: An Industry Analysis*, Occasional Paper 2, Centre for Tourism, Sheffield Hallam University, Sheffield.

Lovelock, J.E. (1979) *Gaia: A New Look at Life on Earth*, Oxford University Press, Oxford.

Lundberg, D.E., Stavenca, M.K. and Krishanmoorty, M. (1995) *Tourism Economics*, John Wiley & Sons, Chichester.

MacCannell, D. (1976) *The Tourist: A New Theory of the Leisure Class*, Macmillan Press, London.

MacCannell, D. (1989) *The Tourist: A New Theory of the Leisure Class*, 2nd edn, Macmillan Press, London.

MacCannell, D. (1992) *Empty Meeting Grounds*, Routledge, London.

MacCannell, D. (2001) 'Remarks on the commodification of culture' in Smith, V.L. and Brent, M. (eds) *Hosts and Guests Revisited: Tourism Issues of the 21st Century*, Cognizant Communications Corporation, New York, 380–390.

McCarthy, M. (2003) 'Death knell for the Kyoto treaty', *The Independent*, London, 3 December: 1.

McLeish, K. (1993) *Key Ideas in Human Thought*, Bloomsbury, London.

Makuni, T.L. (2001) 'The marketing of tourism for ethnic minorities', *Travel and Tourism Analyst*, (1): 77–89.

Malim, T. and Birch, A. (1992) *Social Psychology*, Macmillan Press, Basingstoke.

Malim, T. and Birch, A. (1998) *Introductory Psychology*, Macmillan Press, Basingstoke.

Malone, P. (1998) 'Pollution battle takes to the skies', *The Observer*, London, 8 November: 10.

Mankils, N.G. (2001) *Principles of Economics*, 2nd edn, Harcourt College Publishers, Fort Worth.

Mansfield, Y. (1992) 'From motivation to actual travel', *Annals of Tourism Research*, 19: 399–419.

Maslow, A.H. (1954) *Motivation and Personality*, Harper, New York.

Mason, P. (2003) *Tourism Impacts, Planning and Management*, Butterworth-Heinemann, Oxford.

Mathieson, A. and Wall, G. (1982) *Tourism: Economic, Physical and Social Impacts*, Longman, Harlow.

Mieczkowski, Z. (1995) *Environmental Issues of Tourism and Recreation*, University Press of America, Lanham.

Mill, R.C. and Morrison, A.M. (1992) *The Tourism System: An Introductory Text*, 2nd edn, Prentice Hall, Upper Saddle River.

Mill, R.C. and Morrison, A.M. (2002) *The Tourism System*, 4th edn, Kendall/Hunt Publishing Company, Dubuque.

Milne, S. (1988) *Pacific Tourism: Environmental Impacts and their Management*, Paper presented to the Pacific Environmental Conference, London, 3–5 October.

Mintel (1996) *Snowsports*, Mintel Market Research, London.

Mishan, E.J. (1967) *The Costs of Economic Growth*, Penguin, Harmondsworth.

Momsen, J.H. (1994) 'Tourism, gender and development in the Caribbean' in Kinnaird, V. and Hall, D. (1994) (eds) *Tourism: A Gender Analysis*, John Wiley & Sons, Chichester, 106–120.

Monaghan, J. and Just, P. (2000) *Social and Cultural Anthropology: A Very Short Introduction*, Oxford University Press, Oxford.

Monbiot, G. (1995) 'No man's land', *Tourism in Focus*, 15: 10–11, Tourism Concern, London.

Moscardo, G. (1999) *Making Visitors Mindful: Principles for Creating Sustainable Visitor Experiences through Effective Communication*, Sagamore Publishing, Champaign.

Mowforth, M. and Munt, I. (1998) *Tourism and Sustainability: New Tourism in the Third World*, Routledge, London.

Mowforth, M. and Munt, I. (2003) *Tourism and Sustainability: Development and New Tourism in the Third World*, 2nd edn, Routledge, London.

Naess, A. (1973) 'The shallow and the deep, long-range ecological movement', *Inquiry*, 16: 95–100.

Nash, D. (1979) 'The rise and fall of an aristocratic tourist culture', *Annals of Tourism Research*, Jan./March: 63–75.

Nash, D. (1989) 'Tourism as a form of imperialism' in Smith, V.L. (ed.) *Hosts and Guests: The Anthropology of Tourism*, 2nd edn, University of Pennsylvania, Philadelphia, 36–52.

Nash, D. (1996) *Anthropology of Tourism*, Pergamon, Oxford.

Neal, S. (2002) 'Rural landscapes, representations and racism: examining multicultural citizenship and policy making in the English countryside', *Ethnic and Racial Studies*, 25(3): 442–461.

Nelissen, N., Van der Straaten, J. and Klinkers, L. (1997) *Classics in Environmental Studies: An Overview of Classic Texts in Environmental Studies*, International Book, The Hague.

New Internationalist (1999) 'The Radical Twentieth Century', No. 309, Oxford.

Okasha, S. (2002) *Philosophy of Science*, Oxford University Press, Oxford.

Osborne, R. (1992) *Philosophy for Beginners*, Writing and Readers Publishing, New York.

Ostrom, E., Burger, J., Field, C., Norgaard, R. and Policansky, D. (1999) 'Revisiting the commons: local lessons, global challenges', *Science*, 284: 278–282.

Page, J. (1999) *Travel and Tourism*, *The Guardian*, London, 12 November: 102.

Page, S. (1995) *Urban Tourism*, Routledge, London.

Panos (1995) 'Ecotourism: paradise gained, or paradise lost', *Panos Media Briefing*, 14: 1–15.

Parker, S. (1983) *Leisure and Work*, George Allen and Unwin, London.

Parrinello, G.R. (1993) 'Motivation and anticipation in post-industrial tourism', *Annals of Tourism Research* 20(2): 233–249.

Pass, C., Lowes, B. and Davies, L. (2000) *Dictionary of Economics*, 3rd edn, HarperCollins, Glasgow.

Pattullo, P. (1996) *Last Resorts: The Cost of Tourism in the Caribbean*, Cassell, London.

Pearce, D. (1995) *Tourism Today: A Geographical Analysis*, 2nd edn, Longman, Harlow.

Pearce, P. (1993) 'Fundamentals of tourist motivation' in Pearce, D.G. and Butler, R.W. (eds) *Tourism Research: Critiques and Challenges*, Routledge, London, 113–134.

Pearce, P.L. (1988) *The Ulysses Factor: Evaluating Visitors in Tourist Settings*, Springer-Verlag, New York.

Pearce, P.L. and Stringer, P.F. (1991) 'Psychology and tourism', *Annals of Tourism Research*, 18(1): 136–154.

Pepper, D. (1996) *Modern Environmentalism: An Introduction*, Routledge, London.

Pernatta, J. (1994) 'Corals' in Hare, T. (ed.) *World's Natural Habitats*, Duncan Baird Publications, London, 130–137.

Pfaffenberger, B. (1983) 'Serious pilgrims and frivolous tourists', *Annals of Tourism Research*, 10(1): 57–74.

Phillimore, J. and Goodson, L. (eds) (2004) *Qualitative Research in Tourism: Ontologies, Epistemologies and Methodologies*, Routledge, London.

Plog, S. (1973) 'Why destination areas rise and fall', *Cornell Hotel and Restaurant Administration Quarterly*, February: 55–58.

Plog, S. (1994) 'Developing and using psychographics in tourism research' in Brent Ritchie, J.R. and Goeldner, C.R. (eds) *Travel Tourism and Hospitality Research*, 2nd edn, John Wiley & Sons, New York, 432–444.

Poon, A. (1993) *Tourism, Technology and Competitive Strategies*, CAB International, Wallingford.

Proshansky, H.M. (1973) 'The environmental crisis in human dignity', *Journal of Social Issues*, 29: 1–20.

Prosser, B. (1999) 'Societal change and the growth in alternative tourism' in Cater, E. and Lowman, G. (eds) *Ecotourism: A Sustainable Option?*, John Wiley & Sons, Chichester, 19–39.

Ransom, D. (2001) *The No-Nonsense Guide to Fair Trade*, Verso, London.

Redfern, C. (2003) 'Discover weapons of mass distraction', *The Guardian*, London, 31 July: 6.

Reid, D. (1995) *Sustainable Development: An Introductory Guide*, Earthscan, London.

Retour (2004) 'The case of the Maasai in Loliondo', www.retour.net.

Richardson, D. (1997) 'The politics of sustainable development' in Baker, S., Kousis, M., Richardson, D. and Young, S. (eds) *The Politics of Sustainable Development: Theory, Policy and Practice within the European Union*, Routledge, London, 43–60.

Richter, L. (1995) 'Gender and race: neglected variables in tourism research' in Butler, R. and Pearce, D. (eds) *Change in Tourism's People, Places, Processes*, Routledge, London, 71–91.

Robinson, D. and Garratt, C. (1999) *Introducing Ethics*, Icon Books, Cambridge.

Robinson, M. (1999) 'Cultural conflicts in tourism: inevitability and inequality' in Robinson, M. and Boniface, P. (eds) *Tourism and Cultural Conflicts*, CAB International, Wallingford.

Ross, G.F. (1994) *The Psychology of Tourism*, Hospitality Press, Melbourne.

Ross, G.F. (1998) *The Psychology of Tourism*, 2nd edn, Hospitality Press, Elsternwick, Victoria.

Rostow, W.W. (1971) *The Stages of Economic Growth: A Non-Communist Manifesto*, 2nd edn, Cambridge University Press, Cambridge.

Ryan, C. (ed.) (1997) '*The Tourist Experience: A New Introduction*', Cassell, London.

Sage, A. (2004) 'Fading Côte d'Azur wishes you were here', *The Times*, London, 5 August: 35.

Salem, N. (1995) 'Water rights', *Tourism in Focus*, 17: 4–5.

Sanchez Taylor, J. and O'Connell Davidson, J. (1998) 'Doing the hustle', *Tourism in Focus*, 30: 7–8.

Scheyvens, R. (2002) *Tourism and Development: Empowering Communities*, Pearson Education, Harlow.

Scott, J. (2001) 'Gender and sustainability in Mediterranean island tourism' in Ioannides, Y., Apostolopoulos, E. and Sonninez, E. (eds) *Mediterranean Islands and Sustainable Tourism Development – Practices, Management and Policy*, Continuum, London, 87–107.

Scottish Office (1996) *National Planning Policy Guidelines for Skiing*, Scottish Office, Edinburgh.

Seaton, A.V. (1992) 'Social stratification in tourism choice and experience since the war', *Tourism Management*, 13(1): 106–111.

Seekings, K. (1993) *Politics of Tourism*, Tourism International, London.

Selwyn, T. (1994) 'The anthropology of tourism: reflections on the state of the art' in Seaton, A.V., Jenkins, C.L., Wood, R.C., Dieke, P.U.C., Bennett, M.M., MacLellan, L.R. and Smith, R. (eds) *Tourism: The State of the Art*, John Wiley & Sons, Chichester, 729–736.

Selwyn, T. (ed.) (1996) *The Tourist Image: Myths and Myth Making in Tourism*, John Wiley & Sons, Chichester.

Sharpley, R. (1994) *Tourism, Tourists and Society*, Elm Publications, Huntingdon.

Sharpley, R. (1999) *Tourism, Tourists and Society*, 2nd edn, Elm Publications, Huntingdon.

Sharpley, R. and Telfer, D. (eds) (2002) *Tourism and Development: Concepts and Issues*, Channel View Publications, Clevedon.

Shaw. G. and Williams, A.M. (2002) *Critical Issues in Tourism: A Geographical Perspective*, 2nd edn, Blackwell, London.

Shaw, G. and Williams, A.M. (2004) *Tourism and Tourism Spaces*, Sage, London.

Siegle, M. and Templeton, T. (2004) '30 ways to do the right thing', *The Observer Magazine*, London, 7 November: 34–41.

Simmel, G. (1903) 'The Metropolis and Mental Life', reprinted in Levine, D. (1971) *On Individuality and Social Form*, University of Chicago Press, Chicago.

Simons, P. (1988) 'Après ski le deluge', *New Scientist*, 1: 46–9.

Sinclair, T.M. (1991) 'The tourism industry and foreign exchange leakages in a developing country: the distribution of earnings from safari and beach tourism in Kenya', in Sinclair, T.M. and Stabler, M.J. (eds) *The Tourism Industry: An International Analysis*, CAB International, Wallingford, 185–204.

Sinclair, T.M. (1997) *Gender, Work and Tourism*, Routledge, London.

Sinclair, T.M. and Stabler, M.J. (1997) *The Economics of Tourism*, Routledge, London.

Singer, P. (1993) *Practical Ethics*, 2nd edn, Cambridge University Press, Cambridge.

Slattery, M. (1991) *Key Ideas in Sociology*, Nelson, London.

Smith, C. and Jenner, P. (1992) 'The leakage of foreign exchange earnings from tourism' in *Travel and Tourism Analyst*, (3), Economist Intelligence Unit, London: 52–66.

Smith, M. and Duffy, R. (2003) *The Ethics of Tourism Development*, Routledge, London.

Smith, M.K. (2003) *Issues in Cultural Tourism Studies*, Routledge, London.

Smith, V. and Hughes, H. (1999) 'Disadvantaged families and the meaning of the holiday', *International Journal of Tourism Research*, 1(1): 123–33.

Smith, V.L. (ed.) (1989) *Hosts and Guests: The Anthropology of Tourism*, 2nd edn, University of Pennsylvania, Philadelphia.

Smith,V.L. and Brent, M. (eds) (2001) *Hosts and Guests Revisited: Tourism Issues of the 21st Century*, Cognizant Communications Corporation, New York.

Smout, C. (1990) *The Highlands and the Roots of the Green Consciousness*, Scottish National Heritage, Perth.

Soane, J.V.N. (1993) *Fashionable Resort Regions: Their Evolution and Transformation*, CAB International, Wallingford.

Sofield, T., Bauer, J., De Lacy, T., Lipman, G. and Daugherty, S. (2004) *Sustainable Tourism-Eliminating Poverty (ST-EP): an Overview*, Summary Sheet, World Tourism Organisation, Madrid.

Somerville, C., Rickmers, R.W. and Richardson, E.C. (1907) *Ski Running*, 2nd edn, Horace Cox, London.

Spowers, R. (2002) *Rising Tides: The History and Future of the Environmental Movement*, Canongate, Edinburgh.

Srisang, K. (1991) *The Problematique of Tourism: A View from Below*, Paper presented to the ASEAUK Conference on Tourist Development in South-East Asia, 25–28 March, University of Hull, Hull.

Stephenson, M.L. (2002) 'Travelling to the ancestral homelands: the aspirations and experiences of a UK Caribbean community', *Current Issues in Tourism*, 5(5): 378–425.

Stephenson, M.L. (2004) 'Tourism, racism and the UK Afro-Caribbean diaspora' in Coles, T. and Timothy, D.J. (eds) *Tourism, Diaspora and Space*, Routledge, London, 62–77.

Stokes, P. (2002) *Philosophy: 100 Essential Thinkers*, Capella, London.

Summerskill, B. (2001) 'Package holiday giant dares to think pink', *The Observer*, London, 25 March, http://observer.guardian.co.uk.

Svoronou, E. and Holden, A. (2004) 'Ecotourism as a tool for nature conservation: the role of WWF Greece in the Dadia-Lefkimi-Soufli Forest Reserve in Greece', *Journal of Sustainable Tourism*.

Telfer, D. (2002) 'Tourism development' in Sharpley, R. and Telfer, D. (eds) *Tourism and Development: Concepts and Issues*, Channel View Publications, Clevedon, 1–34.

Thompson, F.M.L. (1988) *The Rise of Respectable Society: A Social History of Victorian Britain 1830–1900*, Fontana, London.

Thompson, M. (1999) *Ethical Theory*, Hodder and Stoughton, London.

TOI (2002) *A Practical Guide to Good Practice: Managing Environmental and Social Issues in the Accommodation Sector*, http: www.toinitiative.org.

Tong, R. (1989) *Feminist Thought: A Comprehensive Introduction*, Routledge, London.

Towner, J. (1996) *An Historical Geography of Recreation and Tourism in the Western World: 1540–1940*, John Wiley & Sons, Chichester.

Towner, J. and Wall, G. (1991) 'History and tourism', *Annals of Tourism Research*, 18(1): 71–84.

Tresidder, R. (1999) 'Tourism in sacred landscapes' in Crouch, D. (ed.) *Leisure/Tourism Geographies: Practices and Geographical Knowledge*, Routledge, London, 137–148.

Tribe, J. (1997) 'The indiscipline of tourism', *Annals of Tourism Research*, 24(3): 638–657.

Tribe, J. (1999) *The Economics of Leisure and Tourism*, 2nd edn, Butterworth Heinemann, Oxford.

Tribe, J. (2004) 'Knowing about tourism: epistemological issues' in Phillimore, J. and Goodson, L. (eds) *Qualitative Research in Tourism: Ontologies, Epistemologies and Methodologies*, Routledge, London, 46–62.

Trigg, R. (1985) *Understanding Social Science*, Blackwell, Oxford.

Tsartas, P., *et al.* (1995) *Oi Koinonikes Epiptoseis tou Tourismou stous Nomous Kerkyras kai Lasithiou* (The social impacts of tourism in the prefectures of Corfou and Lasithi), National Social Research Centre of the Greek National Tourism Board, Athens.

Turner, L. and Ash, J. (1975) *The Golden Hordes: International Tourism and the Pleasure Periphery*, Constable, London.

Turner, V. and Turner, E. (1978) *Image and Pilgrimage in Christian Culture*, Colombia University Press, New York.

UN (2002) *Report of the World Summit on Sustainable Development*, United Nations, New York, 33.

UNEP (2002) *Tour Operators Initiative for Sustainable Tourism Development*, United Nations Environment Program, Paris.

UNEP (2004) *Economic Impacts of Tourism*, http: //www.unepie.org/pc/tourism.

Urry, J. (1990) *The Tourist Gaze: Leisure and Travel in Contemporary Societies*, Sage, London.

Urry, J. (1995) *Consuming Places*, Routledge, London.

Van den Bergh, P.L (1994) *The Quest for the Other: Ethnic Tourism in San Cristobal, Mexico*, University of Washington Press, Seattle.

Veblen, T. (1899) *The Theory of the Leisure Class*, re-published (1994) 3rd edn, Penguin, London.

Vidal, J. (1994) 'Money for old hope', *The Guardian*, London, 7 January: 14–15.

Walker, J. (2000) *Environmental Ethics*, Hodder and Stoughton, London.

Wall, G. (1997) in Pigram, J. and Wahab, S. (eds) *Tourism, Development and Growth*, Routledge, London, 33–49.

Walls, L.D. (2001) 'Henry David Thoreau' in Palmer, J.A. (ed.) *Fifty Key Thinkers on the Environment*, Routledge, London, 106–112.

Walton, J.K. (2002) 'British tourism between industrialisation and globalisation – an overview' in Berghoff, H., Korte, B., Schneider, R. and Harvie, C. (eds) *The Making of Modern Tourism: The Cultural History of the British Experience, 1600–2000*, Palgrave, Basingstoke, 109–133.

Want, P. (2002) 'Trouble in paradise: homophobia and resistance to gay tourism' in Clift, S., Luongo, H. and Callister, C. (eds) *Gay Tourism: Culture, Identity and Sex*, Continuum, London, 191–213.

Washburne, J. (1978) 'Black under participation in wild land recreation: alternative explanations', *Leisure Sciences*, 1: 175–189.

Watson, G.L. and Kopachevsky, J.P. (1996) 'Interpretations of tourism as a commodity' in Apostolopoulus, Y. and Leivadi, S. and Yiannakis, A. (eds) *The Sociology of Tourism: Theoretical and Empirical Investigations,* Routledge, London, 281–297.

Wazir, B. (2001) 'British Muslims fly into a hostile climate', *The Observer*, London, 8 July: 21.

WCED (1987) *Our Common Future*, Oxford University Press, Oxford.

Wheat, S. (1999a) 'Editorial of tourism on trial', *Tourism in Focus*, 33: 3.

Wheat, S. (1999b) 'To go or not to go to Indonesia', *Tourism in Focus*, 33: 5.

Wheeller, B. (1993) 'Sustaining the ego', *Journal of Sustainable Tourism*, 1(2): 23–29.

Whitelegg, J. (1999) *Air Transport and Global Warming*, www.gn.apc.org/sgr/kyoto/jw.html.

Wigg, J. (1996) *Bon Voyage: Travel Posters of the Edwardian Era*, HMSO, London.

Wight, P. (1994) 'Environmentally responsible marketing of tourism' in Cater, E. and Lowman, G. (eds) *Ecotourism: A Sustainable Option*, John Wiley & Sons, Chichester, 39–53.

Williams, A. (1997) 'Tourism and uneven development in the Mediterranean' in King, R., Proudfoot, L., and Smith, B. (eds) *The Mediterranean: Environment and Society*, Arnold, London, 208–225.

Williams, B. (2000) 'Plato: the invention of philosophy' in Monk, R. and Raphael, F. (eds) *The Great Philosophers*, Phoenix, London, 47–92.

Witt, A.C. and Wright, L.P. (1992) 'Tourist motivation: life after Maslow' in Johnson, P. and Thomas, B. (eds) *Choice and Demand in Tourism*, Mansell, London, 33–55.

World Bank (2000) *Entering the 21st Century: World Development Report 1999/2000*, Oxford University Press, Oxford.

WTO (1992) *Tourism Carrying Capacity: Report on the Senior-level Expert Group Meeting Held in Paris, June 1990*, World Tourism Organisation, Madrid.

WTO (1998) *Tourism Economic Report: 1st Edition*, World Tourism Organisation, Madrid.

WTO (1999) *Global Code of Ethics for Tourism*, World Tourism Organisation, Madrid.

WTO (2002) *Tourism and Poverty Alleviation*, World Tourism Organisation, Madrid.

WTO (2003a) *Tourism Highlights (Edition 2003)*, World Tourism Organisation, Madrid.

WTO (2003b) *Compendium of Statistics*, World Tourism Organisation, Madrid.

WTO (2003c) *Climate Change and Tourism*, World Tourism Organisation, Madrid.

WTO (2004a) *Tourism and World Economy*, www.world-tourism.org/facts/trends/economy. htm.

WTO (2004b) *World's Top Tourism Spenders*, www.world-tourism.org/facts/trends/ expenditure.htm.

WTO (2004c) '*Methodological Notes*, www.world-tourism.org/facts/methodological/ method.htm.

WTO (2004d) *International Tourism Receipts by Country of Destination*, www.world-tourism.org/facts/trends/receipts.htm.

WTTC (2002) *Corporate Social Leadership in Travel and Tourism*, World Travel and Tourism Council, London.

WTTC (2004) *World Travel and Tourism Forging Ahead*, World Travel and Tourism Council, London.

Yiannakis, A. and Gibson, H. (1992) 'Roles tourists play', *Annals of Tourism Research*, 19: 287–303.

Youell, R. (1998) *Tourism: An Introduction*, Longman, Harlow.

Zarkai, C. (1996) '*Philoxenia*: receiving tourists – but not guests – on a Greek island' in Boissevain, J. (ed.) (1996) *Coping with Tourists: European Reactions to Mass Tourism*, Berghahn Books, Oxford, 143–173.

Zimmermann, F.M. (1995) 'The alpine region: regional restructuring opportunities and constraints in a fragile environment', in Montanari, A. and Williams, A.M. (eds) *European Tourism: Regions, Spaces and Restructuring*, John Wiley & Sons, Chichester.

Index